PETROLEUM CHEMICALS - RECENT INSIGHT

Edited by **Mansoor Zoveidavianpoor**

Petroleum Chemicals - Recent Insight

http://dx.doi.org/10.5772/intechopen.73337

Edited by Mansoor Zoveidavianpoor

Contributors

Adeyinka Yusuff, Olalekan Adeniyi, Moses Aderemi Olutoye, Uduak George Akpan, Cheng Seong Khor, Shuai Wang, Nirmal Nirmal Ghimire, Constantinos Tsanaktsidis, Konstantinos Spinthiropoulos, Mansoor Zoveidavianpoor

Notice

Statements and opinions expressed in the chapters are these of the individual contributors and not necessarily those of the editors or publisher. No responsibility is accepted for the accuracy of information contained in the published chapters. The publisher assumes no responsibility for any damage or injury to persons or property arising out of the use of any materials, instructions, methods or ideas contained in the book.

First published in London, United Kingdom, 2019 by IntechOpen

IntechOpen is the global imprint of INTECHOPEN LIMITED, registered in England and Wales, registration number: 11086078, The Shard, 25th floor, 32 London Bridge Street
London, SE19SG – United Kingdom
Printed in Croatia

British Library Cataloguing-in-Publication Data
A catalogue record for this book is available from the British Library

Additional hard and PDF copies can be obtained from orders@intechopen.com

Petroleum Chemicals - Recent Insight, Edited by Mansoor Zoveidavianpoor
p. cm.
Print ISBN 978-1-83880-401-5
Online ISBN 978-1-83880-402-2
eBook (PDF) ISBN 978-1-83880-652-1

We are IntechOpen,
the world's leading publisher of
Open Access books
Built by scientists, for scientists

4,100+
Open access books available

116,000+
International authors and editors

125M+
Downloads

Our authors are among the

151
Countries delivered to

Top 1%
most cited scientists

12.2%
Contributors from top 500 universities

Interested in publishing with us?
Contact book.department@intechopen.com

Numbers displayed above are based on latest data collected.
For more information visit www.intechopen.com

Meet the editor

Dr. Mansoor Zoveidavianpoor has over 18 years of multidisciplinary oil and gas experience, built on his technical, operational, and management roles in the industry and academia. Dr. Zoveidavianpoor holds a PhD degree in Petroleum Engineering from the University of Technology Malaysia (UTM). He was involved in different disciplines such as geology, flow assurance, piping construction, artificial intelligence, environmental engineering, petroleum engineering, and project management. He has lectured several courses at UTM, the Petroleum University of Technology, and Islamic Azad University. He is a member of the Society of Petroleum Engineers and is registered as a chartered petroleum engineer. He has published more than 50 publications in international peer-reviewed journals and conferences, has contributed to five textbooks, and has served on many worldwide scientific committees. Previously, he was working as a senior lecturer at UTM and as a senior petroleum engineer at the National Iranian Oil Company. Currently, he works as a production technologist at PETRONAS. Dr. Zoveidavianpoor is actively involved in multidisciplinary studies and currently his main area of focus is on unconventional reservoir management.

Contents

Preface

The book aims to add contributions and new advances in technologies and treatment on petroleum chemicals, in terms of oilfield chemicals, biofuel production, and chemical transformation. The book begins with an introduction on oilfield chemicals.

Chemicals have been an essential part of the petroleum industry in the current and past centuries. An essential for modern industry is to have access to low-cost energy. Worldwide, petroleum consumption will reach 100 million barrels per day, more than twice it was 50 years ago. Nevertheless, petroleum resources are declining, and overwhelming greenhouse gases, by releasing 400 billion tons of carbon, threaten the Earth. The fact is preventing cars and trucks from using petroleum fuels is a difficult task. Accordingly, we need to cure our dependence on petroleum and be able to develop sustainable solutions to fuel our future. The challenges associated with sustainable hydrocarbon production have seen a major growth in the need for biofuel and chemical transformations. A promising solution involves biodiesel from natural oils and fats (mineral oils), which could be used to provide heat and electricity. By blending waste frying oil with petroleum diesel, scientists have devised a new way to convert waste cooking oil into biodiesel that could make it more affordable, as shown in the chapter "Waste Frying Oil as a Feedstock for Biodiesel Production."

The global drive for environmental sustainability necessitates continuous adjustment, optimization, and improvement in petroleum refining processes to generate energy and products that include automotive fuels such as gasoline. In gasoline engines, knocking results in destructive effects to the engine and drastically increases the pressure inside the engine's cylinders. So, clearly, antiknocking agents are additives that prevent or reduce engines from knocking. Catalytic reforming is an important process in the petroleum refining industry, which developed originally to produce components of automotive fuels, specifically gasoline, which meet engine requirements for high antiknock quality. In fact, refiners need to maximize their asset utilization to maintain competitiveness in the business setting. As shown in the chapter "High-Octane Gasoline Production from Catalytic Naphtha Reforming," utilizing such an application will contribute toward the production of energy in an environmentally sustainable manner through an optimal process operation approach with reduced off-specification fuel products.

In the next chapter, "Implementation of Basic Principles of Econometric Analysis in Petroleum Technology: A Review of the Econometric Evidence," a question on how the physicochemical parameters of distilling petroleum products can be understood is answered for practitioners. In this review chapter, the quality of the different fuels is expressed by a series of physical, chemical, and other characteristics. The connection between production process and quality of fuel is crucial in the field of petroleum technology. Results show that the regression analysis perfectly illustrates the relationships between the variables in all applied models.

Rising global oil consumption in modern society has led to more petroleum waste generation. Petroleum waste is full of pollutants and its treatment aims at reducing the contaminants to acceptable levels to make the water safe for discharge back into the environment. Having high concentrations of aliphatic, aromatic petroleum hydrocarbons, oil processing wastewater will affect plants and aquatic life of surface and groundwater sources. Due to its organic origination, complex nature, and toxic effects, wastewater treatment prior to discharge is obligatory. The biological treatment process is normally applied to reduce the effects of petrochemical waste. Powered by two case studies, as shown in the chapter "Biological Treatment of Petrochemical Wastewater," the commonly applied pretreatment methods for petrochemical wastewater are summarized and compared with biological treatment performance of different systems.

Mansoor Zoveidavianpoor, PhD
MEI Chartered Petroleum Engineer
Executive Production Technology
Petronas
Kuala Lumpur, Malaysia

Chapter 1

Introductory Chapter: Oil Field Chemicals - Ingredients in Petroleum Industry

Mansoor Zoveidavianpoor

Additional information is available at the end of the chapter

http://dx.doi.org/10.5772/intechopen.85957

Time is the friend of the wonderful business, the enemy of the mediocre - Alice Schroeder

1. Introduction

The real task for the oil industry is how quickly it can move to take advantage of the many opportunities that "gas and renewable" technologies will bring. While oil demand slowly falls with the adoption of more renewables and gas technologies, there is a need for oil companies to have insight into new technology advancement and accordingly innovate to stay competitive and keep the fuel flowing. A great deal of activities in the oil and gas sector is focused on upstream and downstream, and not surprisingly, research and development still plays a key role in the coming years. Oil companies should be prepared to pursue new drilling and extraction technologies and to increase their research into sustainability and clean energy. I think oil sector leaders might consider a question on how their companies can develop new capabilities and in what areas?

Nowadays, improved oil recovery (IOR) is one of the main strategic priority areas in petroleum industry [1]. IOR processes consist of all techniques that are employed to enhance hydrocarbon production. Oil field chemicals have many positive functions such improved oil recovery, drilling optimization, corrosion protection, prevent mud loss in different geological formations, stabilize drilling fluid in high pressure and high temperature environment, and many others [2]. Oil field chemicals demand is expected to reach USD 32.69 billion by 2023 from USD 26.06 billion in 2017 [3]. The rising demand from Asia-Pacific, shale gas, and increasing deep water drilling operations are likely to be the major driven for the oil field chemicals market.

Region wise, oil field chemicals have received much attention in recent years (**Figure 1**) due to their contribution in oil recovery of hydrocarbons, which offer important economic benefits. Many case studies and lessons learned from the industry show that there are excellent

- North America
- Latin America
- Middle East
- CIS
- Africa
- Europe
- Asia Pacific
- China

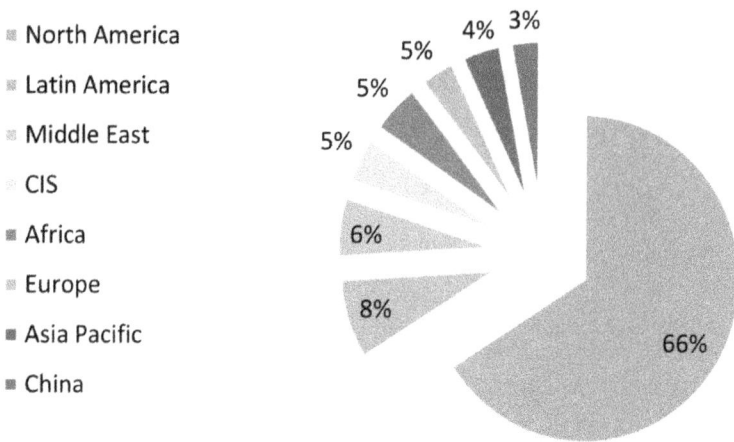

Figure 1. Oil field chemicals by region.

- Stimulation Chemicals
- Drilling Fluids
- Production Chemicals
- EOR Product
- Workover Fluids
- Cement Aditives

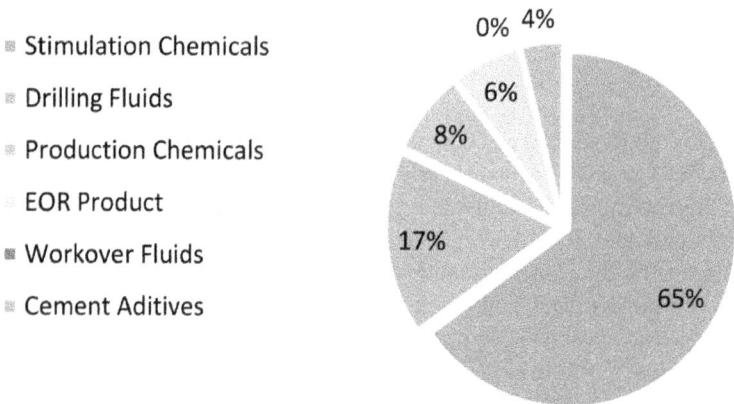

Figure 2. Oil field chemicals market.

opportunities to grow for oil field chemicals in certain fields such as drilling and cementing, enhanced oil recovery, production, workover and completion, and well stimulation. Well stimulation comprises of different types of operations performed on a well to maintain and/ or improve its productivity [4].

According to IHS Markit, 2018 [5], oil field chemicals enable the production of oil and gas or make it efficient and is projected to grow at an average annual rate of about 4% during 2017–2022. Logistics (hauling, transfer, and storage) and disposal issues are directly related to the green and continuous development in oil field chemicals. These two are contributed to approximately 85–90% of total annual spending money in petroleum industry.

The wide range of oil field chemicals, including well stimulation and other additives, plays an important role in maximizing the productivity of existing (green) and mature (brown) fields.

Stimulation operations can be conducted solely on the wellbore or on the reservoir; those can be performed on old wells and new wells alike; and it can be designed for remedial purposes or for enhanced production. As shown in **Figure 2**, the market size of the well stimulation in USA alone is about 61% of the total oil field chemicals. Increases in well stimulation activities are expected to continue; accordingly, development and innovation in stimulation chemicals will effectively shape the future of oil industry; that is one of the potentials what the oil and gas leaders may think about!

I am calling oil field chemicals as "every time ingredients" in petroleum industry; they are utilized in many ways and different stages in downstream and upstream sectors, starting from drilling, production, stimulation, and finally abandonment. So, if the oil sector leaders are thinking about innovative solutions in oil field chemicals, TIME is really their friend.

Author details

Mansoor Zoveidavianpoor

Address all correspondence to: mansoor353@yahoo.com;
mansoor.zoveidavian@petronas.com.my

Executive Production Technology, PETRONAS, Kuala Lumpur, Malaysia

References

[1] Zoveidavianpoor M, Shadizadeh SR, Mombeini S. Evaluation and improvement of well stimulation by matrix acidizing in one of the Southern Iranian oilfields. Petroleum Research. 2010;**20**(62):94-106

[2] Fink J. Petroleum Engineer's Guide to Oil Field Chemicals and Fluids. Waltham, MA, USA: Gulf Professional Publishing; 2015

[3] Freedonia Group. Oil Field Chemicals. 2015. Available from: https://www.freedonia-group.com/industry-study/oilfield-chemicals-3352.htm

[4] Gharibi A, Zoveidavianpoor M, Daraei Ghadikolaei F. On the application of well stimulation method in improvement of oil recovery. Applied Mechanics and Materials. 2015;**735**:31-35. DOI: 10.4028/www.scientific.net/AMM.735.31

[5] IHS Markit. Oil Field Chemicals. 2018. Available from: https://ihsmarkit.com/products/chemical-oil-field-scup.html

Chapter 2

Waste Frying Oil as a Feedstock for Biodiesel Production

Adeyinka Sikiru Yusuff, Olalekan David Adeniyi,
Moses Aderemi Olutoye and Uduak George Akpan

Additional information is available at the end of the chapter

http://dx.doi.org/10.5772/intechopen.79433

Abstract

This study was initiated to blend the biodiesel produced from waste frying oil (WFO) with petroleum diesel in three different proportions (B20, B50 and B80), and the dual fuels were tested on compression ignition engine to evaluate their emission characteristics. The biodiesel produced from WFO was achieved via heterogeneous catalyzed transesterification using anthill-eggshell-Ni-Co mixed oxide composite catalyst at reaction temperature of 70°C, reaction time of 2 h, catalyst loading of 3 wt% and methanol to oil molar ratio of 12:1. Various analyses carried out on the prepared WFO-based biodiesel confirmed that it is of good quality and also meet the ASTM standard. The blended fuel containing 20% by volume biodiesel content (B20) emitted 1050 and 14,000 ppm of CO and CO_2, respectively, which were lower than those emitted by B0, B50 and B80. It can be concluded that blending the diesel with appropriate volume of biodiesel not only improves its quality but also lowers greenhouse gases emission.

Keywords: biodiesel, catalyst, diesel, transesterification, diesel engine, waste frying oil

1. Introduction

The basic concept of biodiesel synthesis was invented by Rudolf Diesel in the 1890s, and the diesel engine had become the device of choice for power reliability and high fuel economy globally. He envisaged that pure vegetable oil could be used on compression ignition engine as fuel [1]. After it was used, it became obvious that it was not suitable to power diesel engine due to some limitations associated with its use. Those limitations include carbon deposition in heating chamber, sticking of piston ring, injection tip coking and partial combustion [2–4].

However, modern biodiesel fuel has its basis in research conducted in Belgium [5], but bio-diesel production plant was not set up in any of the European countries until the late 1980s [4]. Continents in the other parts of the world including Africa also witness the local production of biodiesel starting up. Today, there are many countries with commercial biodiesel projects [6].

Renewable and alternative fuel such as biodiesel is capable of solving associated problems with fossil hydrocarbon fuel. Greenhouse gas emissions arising as a result of fossil hydro-carbon fuel burning in vehicular or compression ignition (CI) engine have been identified as the main problem confronting the entire world nowadays [7]. Recently, interest in non-toxic, renewable, biodegradable, alternative fuel such as biodiesel and bio-hydrogen, with their unique applications in powering vehicular and CI engines, is ongoing. However, the problems related to sticking of piston ring and injector tip coking do occur in long run usage of biofuel without necessarily adjusting fuel consumption and engine performance [8]. Since carbons present in biodiesel are biodegradable, it contributes less to carbon cycle. Besides, the qualities of petrol-diesel get improved and the emissions of sulphur and nitrogen oxides get reduced when biodiesel is blended with appropriate quantity of diesel [9].

Standard specifications for biodiesel have been adopted by most countries, for instance, America and Europe use ASTM D6751 and EN 14212, respectively. Generally, a code consist-ing of a number that indicates the biodiesel content in percentage is employed, for example, B100 is purely 100% biodiesel while B40 simply means a mixture containing 40% biodiesel and 60% petrol-diesel. In some part of the Europe, Sweden in particular, a dual fuel contain-ing 5% biodiesel (B5) is often used [10]. Recently in Nigeria, the Nigerian National Petroleum Corporation (NNPC) adopted B20 as standard specification for biodiesel-fossil diesel blend. This according to the corporation would require 80 volumes of petrol-diesel to be blended with 20 volumes of biodiesel.

Generally, any material that contains triglyceride can be used to produce biodiesel [11]; however, a choice of feedstock to be used should be carefully made. At present, it is usually made from edible and non-edible plant oils obtained from palm kernel seed [12–14], soybean [15–18], rapeseed [19–22], coconut oil [23], sunflower oil [24–30], Tiger nut [31, 32] (*Cyperus esculentus*), cotton seed [33] and Sorghum [34]. The use of oil from algae [35], fish [36], Karanja [37], *Jatropha curcas* seed [38–41], yellow horn corn oil [42] and Chinese tallow seed oil [43] for the synthesis of biodiesel had also been reported. It has also been reported that biodiesel can be conveniently synthesized from waste vegetable oils [10].

1.1. Waste frying oil as economic feedstock for biodiesel synthesis

Waste frying oil (WFO) is regarded as spent oil which has been employed for deep frying and is no more viable for further consumption. WFO is made up of saturated and unsaturated monocarboxylic acids with the trihydric alcohol glyceride saturated and unsaturated mono-carboxylic acids [44]. During frying, the physical, chemical and organoleptic features of the plant oil change [45]. More so, free fatty acids content are enhanced in the oil by hydrolysis of reactive components as a result of water from food during heating. High cost of biodiesel has been identified as the major reason why its production has not been widely commercialized.

One of the ways in which this could be addressed is to develop a holistic method to minimize the biodiesel cost [46, 47]. Those options include biodiesel synthesis from spent frying oil and also minimize its processing cost through the optimization of process parameters that have influence on its yield and quality [48].

Oils which contain high free fatty acids (FFAs) such as waste cooking oil are now being used for biodiesel synthesis, because they are less expensive than refined oil [49, 50]. More so, it offers significant advantages which include reduction in environmental problem and the production cost [51]. WFO is available in large amount, and its management constitutes a serious disposal problem. In most of the developed countries, spent frying oil is being used as raw material for making soap [52].

According to United States Energy Information Administration Agency, several gallons of used cooking oil are collected daily in United States of America [52], where close to 9 pounds of used cooking oil are produced per person each year [53]. In Europe, about 0.49–0.7 million gallons per day of waste frying oil are collected [54]. In Nigeria, WFO is one of the major wastes generated in hotel, restaurants and eateries [55]. However, since no strict and more stringent environmental legislations on WFO discharge, those organizations mentioned earlier discharge WFO indiscriminately into water bodies and on land, thus leading to environmental degradation [52]. More so, there is no Information Administrative Agency that accounts for WFO generated in Nigeria. However, the operation of Students' Cafetaria at Afe Babalola University (ABUAD), Ado-Ekiti, Nigeria, where the waste frying oil used in this study was collected, is being monitored by food scientist and nutritionist. They make sure that the used oil is not reused more than two times after initial frying to prevent the intake of free fatty acid and thus use the WFO as raw material for producing soap.

Table 1 shows the average fatty acid composition in waste frying oil [50]. However, low quality feedstock, which contains high concentration of free fatty acid (greater than 1%), cannot be

Fatty acid (trivial/rational name)	Methyl ester (trivial/rational name)	Formula	Common acronym	Acid composition (%)
Palmitic acid/hexadecanoic acid	Methyl palmitate/methyl hexadecanoate	$C_{16}H_{32}O_2$	C16:0	15.86
Stearic acid/octadecanoic acid	Methyl stearate/methyl octadecanoate	$C_{18}H_{36}O_2$	C18:0	4.87
Oleic acid/9 (E)—octadecenoic acid	Methyl oleate/methyl 9 (E) octadecenoate	$C_{18}H_{34}O_2$	C18:1 (E)	29.83
Linoleic acid/9 (Z), 12 (Z)—octadecadienoic acid	Methyl linoleate/methyl 9 (Z), 12 (Z) octadecadienoate	$C_{18}H_{30}O_2$	C18:2 (Z,Z)	28.85
Linolenic acid/9 (Z), 12 (Z), 15 (Z)—octadecatrienoic acid	Methyl linoleate/ methyl 9 (Z), 12 (Z), 15 (Z)—octadecadienoate	$C_{18}H_{30}O_2$	C18:3 (Z,Z,Z)	2.49

Source: [50].

Table 1. Average fatty acid composition in waste frying oil.

easily transformed into biodiesel by alkali transesterification because it consumes the catalyst and reduces its performance leading to substantial yield losses [56, 57].

Acid-catalyzed transesterification remains the best means of converting oil with high FFA content to biodiesel, but due to the harsh reaction conditions and prolong time of reaction [42, 46, 58], it has been largely ignored. Talebian-Kiakalaieh et al. [51] investigated that heterogeneous transesterification of WFO rich in FFA with methanol using heteropoly acid (HPA) catalyst and 88.6% of biodiesel was obtained at optimum reaction conditions. Meanwhile, the authors only focused attention on the performance and reusability of the heterogeneous acid catalyst while process economy was not given attention.

Since the transesterification of waste frying oil rich in FFA may not proceed using solid base catalyst, a two-step transesterification reaction is often employed [56, 59, 60]. The first step is the acid esterification process through which high FFA oils react with methanol using mineral acid usually concentrated tetraoxosulphate (VI) acid as catalyst to produce free fatty acid ester [44, 50, 61]. Followed by the basic transesterification process, whereby acid preheated waste frying oil reacts with methanol in the presence of base catalyst to form biodiesel and other products [52]. Therefore, a two-step transesterification process not only removes high free fatty acid content (FFA) but also improves the biodiesel yield [44]. The test of the waste frying oil-derived biodiesel on commercial diesel engines provides better performances and less gaseous pollutant is emitted apart from NO_x [62].

2. Raw materials

Apart from the biodiesel feedstock, alcohol and catalyst also play important roles in biodiesel production. The required raw materials for the synthesis of biodiesel from WFO are explained as follows:

2.1. Alcohol

In biodiesel synthesis, monohydric alcohols, such as methanol, ethanol and butanol, are usually employed as co-reactant. Methanol is a basic monohydric alcohol used in excess for the production of biodiesel via catalyzed transesterification process. Methanol is light, volatile, poisonous and inflammable and contributes to ozone layer depletion. Biomass-derived fuel produced with methanol from natural gas or coal has approximately 94–96% biogenic content [63]. However, a 100% renewable biodiesel could be produced, if bioethanol is used as a substitute for methanol [64]. Ethanol is also a light alcohol, volatile, flammable, colourless and biodegradable. Among those aforementioned alcohols, methanol is mostly used for biodiesel production commercially, because it is relatively cheap, readily available and easier to separate glycerol from the product mixture [65]. The use of ethanol and other monohydric alcohols for biodiesel production has however all been reported [66].

2.2. Catalyst

Industrial production of biodiesel is frequently done via homogeneous catalyzed transesterification process, whereby the triglyceride contained in vegetable oil or animal fat reacts with alcohol (methanol/ethanol) in the presence of liquid catalyst. Utilization of enzymes as biocatalyst for the transesterification of triglyceride to biodiesel has also been reported [67]. However, due to the problem associated with homogeneous catalysis (wastewater generation and difficulty in reusability) and enzymes (exorbitant cost and deactivation), there have been simulated researches in the field of heterogeneous catalysis for biodiesel production [68, 69]. The solid-based catalysts include pure metal oxides, mixed metal oxides, alumina, silica and zeolite-supported catalyst, sulphated metal oxide and those ones derived from waste and naturally occurring materials. Vujicic et al. [70] investigated the transesterification of sunflower oil using CaO catalyst. Apart from that, Wen et al. [43] studied the transesterification reaction between cotton seed oil and methanol using TiO-MgO as heterogeneous catalyst. Many more heterogeneous catalysts suitable for biodiesel production have been reported in literature [23, 71].

3. Materials and methods

3.1. Materials

The WFO after used for long frying was collected from students' Cafeteria 1, Afe Babalola University, Ado-Ekiti, Nigeria. The chemicals/reagents used for waste frying oil characterization such as potassium hydroxide (KOH), ethanol (95%), hydrochloric acid (HCl), diethyl ether, phenolphthalein, diethyl ether, chloroform and acetic acid (BDH, England) were all obtained from Chemical Science Laboratory, ABUAD, Ado-Ekiti, while analytical grade methanol (JHD, AR China) was procured from Nizo Chemical Enterprise, Akure and was used as received. Distilled water was prepared in the laboratory. The heterogeneous catalyst used for this study was a self-synthesized, and it is known as anthill-eggshell-Ni-Co mixed oxide composite catalyst (AENiCo).

3.2. Characterization of waste frying oil

The WFO was first heated in a hot air oven at 120°C for 4 h and later filtered using a 120-μm sieve mesh to remove any non-oil components or bits of food residues. The basic physico-chemical properties of the waste frying oil shown in **Figure 1** are determined as follows:

3.2.1. Density

The empty density bottle was weighed and recorded as w_1. Prior to this, the temperature of the oil was taken with thermometer and obtained to be 23°C. The bottle was filled with distilled water, after which it was weighed and recorded as w_2. Furthermore, the bottle was emptied, cleaned with tissue paper and also filled with equal volume of waste frying oil. The weight of the bottle and waste frying oil was then measured and recorded as w_3. The density of the sample was thereby determined using Eqs. (1) and (2).

$$R.D = \frac{w_0}{w_w} = \frac{w_3 - w_1}{w_2 - w_1} \tag{1}$$

$$\rho_{wfo} = R.D \times 1 \text{ g/cm}^3 \tag{2}$$

Figure 1. Samples of waste frying oil.

3.2.2. Kinematic viscosity

The essence of subjecting vegetable oil (edible and non-edible) to transesterification reaction process is to reduce its viscosity and density, because it cannot be used directly to power any diesel engine. This property needs to be determined before and after biodiesel synthesis from vegetable oil. In order to determine this property, a weighed quantity of oil was poured into a stainless cup of DV-III ultra programmable rheometer (Brookfield: model LVDV-III U) and preheated to a temperature of 60°C for 1 h. Thereafter, its spindle and cup containing heated oil were placed under the rheometer. The rotor and spindle were both immersed in the cup after the rheometer had been set to require speed (150 rpm). The value of dynamic viscosity was immediately taken when the temperature of the preheated oil dropped to 40°C. However, in order to determine the kinematic viscosity, the density of waste frying oil at 40°C was therefore determined using Eq. (3).

$$\mu_K = \frac{\mu_D}{\rho_{wco40}}. \tag{3}$$

3.2.3. Acid value

It is the number of milligram potassium hydroxide needed to neutralize the acid contained in 1 g of oil or fat sample. It measures the extent at which the glyceride contained in an oil sample decomposes by the activity of lipase or other actions. Acid value of the WFO was examined by titration method reported elsewhere [72].

In this method, 5.926 and 6.695 g of oil samples were weighed and poured into flasks A and B, respectively, and 25 mL each of diethyl ether and ethanol was added into flasks A and B and 2 drops of 1% phenolphthalein indicator solution were also added into each of the flasks. Thereafter, potassium hydroxide (KOH) solution was then titrated against dissolved oil-solvent mixture under constant agitation until the solution turned into pink as shown in Plate VI. The acid value (A.V) and free fatty acid content (% FFA) were thus calculated using Eqs. (4) and (5).

$$(A.V) \text{ in mgKOH/g} = \frac{56.1 \times C_{KOH} \times V_{KOH}}{w_o} \qquad (4)$$

$$\%FFA = \frac{A.V}{2} \qquad (5)$$

Where V_{KOH} is the volume of potassium hydroxide, C_{KOH} is the concentration of potassium hydroxide, w_o is the weight of oil used whose value must lie between 0 and 10 g, AV is acid value and % FFA is the percentage of free fatty acid.

3.2.4. Saponification value

The saponification value of an oil or fat is the number of mg of potassium hydroxide required to neutralize the fatty acids resulting from the complete hydrolysis of 1 g of the sample. It helps in detecting oils and fats, which contain a high proportion of the lower fatty acids. The principle is that oil or fat undergoes saponification reaction with large volume of alcoholic potassium hydroxide, and the resulted product is subjected to titration process in order to determine the amount of potassium hydroxide remaining after saponification reaction.

In order to determine the saponification value of WFO, 8.221 g of KOH pellet was weighed and dissolved in 5 ml distilled water, after which 250 ml of 95%v/v ethanol was added and allowed to settle overnight. Decant of the clear solution of alcoholic potassium hydroxide was then obtained. Also, 1 g of phenolphthalein powder was dissolved in 100 ml ethanol to make phenolphthalein indicator solution.

About 2.030 g of oil was weighed into flat bottom flask A and 25 mL of alcoholic potassium hydroxide solution was added into the flask A. Another flat bottom flask B was also filled with 25 mL of alcoholic KOH without oil. The reflux condensers were attached to each of the flasks and heated on a boiling water bath for 1 h with occasional shaking. The flasks A and B were then removed from the water bath after 1 h and two drops of phenolphthalein indicator were added into each of the flasks. Both colours in flasks A and B changed to pink. While still hot, contents in flasks A and B were titrated with the standard 0.5 M hydrochloric acid with constant shaking until the solutions in flask A and B changed back to pale yellow and colourless, respectively. However, after the estimation of volume of acid consumed, the saponification value (SV) was then calculated using Eq. (6) [50].

$$S.V = \frac{(b-a) \times 28.05}{w_o}(\text{mgKOH/g}) \qquad (6)$$

Where a and b are volumes of acid used against alcoholic KOH solution with oil and alcoholic KOH solution without oil, respectively. w_o is the volume of WFO used, while S.V is the saponification value.

3.2.5. Average molecular weight

It is the weight in atomic mass units of all the atoms in a given formula. The determination of molecular weight of a substance is necessary in order to ascertain the number of grams contained in one mole of that same substance. It is a function of saponification and acid values. Average molecular weight (AMV) of oil is determined using Eq. (7).

$$A.M.W = \frac{56.1 \times 1000 \times 3}{(S.V - A.V)} \tag{7}$$

Where A.M.W is the average molecular weight. Saponification and acid values are denoted by S.V and A.V, respectively.

3.3. Biodiesel production from WFO

The transesterification of WFO to biodiesel using AENiCo catalyst was carried out in a batch reactor made up of a 250 mL one way round bottom flask fitted with a condenser and thermometer as shown in **Figure 2**. The reaction was performed at reaction conditions, considering catalyst loading of 3 wt%, reaction temperature of 70°C, reaction time of 2 h and methanol to WFO molar ratio of 12:1, while stirring rate was kept constant throughout the reaction. At the end of the reaction, the resulting mixture was filtered using white cloth in order to remove the spent catalyst, and the filtrate was then poured into a separating funnel and left there overnight to settle. During the process, two layers of liquid were observed, in which the upper layer was biodiesel and the lower layer indicated glycerol.

3.4. Preparation of different blends of biodiesel and petroleum diesel

In this study, waste frying oil-derived biodiesel was mixed with petroleum diesel and used on ignition engine to evaluate its performance and characterize the exhaust gas emission. The mixing of two different diesels (biodiesel of waste frying oil and fossil diesel) was made in plastic container coupled with electric mixing machine as shown in **Figure 3**. The blends were made at different proportions with 20% (B20), 50% (B50), 80% (B80) and 100% (B100) by volume of biodiesel. It was vigorously mixed with the help of agitator being driven by electric motor at room temperature with agitation speed of 400 rpm. The characteristic of the prepared biodiesel, biodiesel-diesel blend and diesel including specific gravity, kinetic viscosity, lower heating value (LHV) and flash point were determined. The LHV of the fuels was determined using Eq. (8) [73].

$$LHV = -0.167\rho + 184.95 \tag{8}$$

3.5. Emission characterization of compression ignition engine

The method reported by Elsolh [74] was adopted for the performance evaluation of the engine. The gas emission of three different blends of biodiesel and diesel were measured and compared in a Yoshita S195NM ignition engine whose technical features are given in **Table 2**. The first blend used was B20, it was poured into the fuel tank of the ignition engine and the engine was immediately turned on by hand whirling. The probe of a gas analyzer was thereafter attached to exhaust pipe of the engine. The engine was left to work for almost 20 min in order to make it stabilize and allow the thick smoke to escape. The measurement was then taken

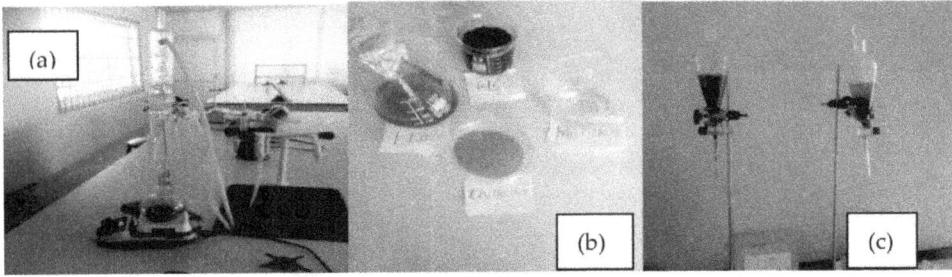

Figure 2. (a) Experimental set, (b) raw materials and synthesized biodiesel and (c) biodiesel-glycerol mixture.

Figure 3. (a) Petroleum diesel and WFO-derived biodiesel, (b) blending of biodiesel and petroleum diesel and (c) blended fuels.

Technical properties	Values
Number of stroke	4
Number of cylinder	1
Declared speed	2000 rpm
Compression ratio	20:1
Rated power	3.32 kW
Overall dimensions	900 × 440 × 760 mm
Bore and stroke	95 × 115 mm

Table 2. Ignition engine specification.

by gas analyzer every 5 min for 20 min, and the values of the emission measurements were stored on an input computer program to determine the average values. The same procedure was used for other fuels (B50 and B80) (**Figure 4**).

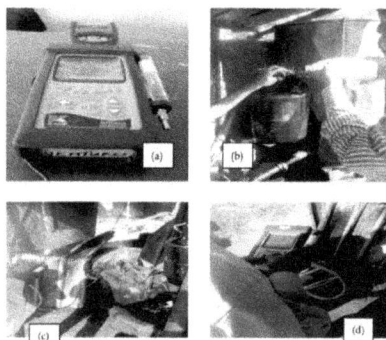

Figure 4. (a) KANE AUTOplus gas analyzer, (b) blended fuel being poured into the fuel tank of diesel engine, (c) gas emission analysis and (d) saving of analysis results on analyzer.

4. Results and analysis

4.1. Characterization of WFO

In this current study, WFO used as feedstock was characterized based on its physicochemical properties. The properties are summarized in **Table 3**. Its density at 25°C was determined to be 0.9147 g/cm³. The obtained value of density was slightly less than those ones recorded by Chhetri et al. [52] and Mahgoulb et al. [75] as 0.9216 and 0.9185 g/cm³ at 23°C, respectively. This difference is attributed to the fact that density is a function of temperature [52] and decreases as temperature increases [76]. The kinematic viscosity of waste frying oil was 9.36 cP.

More so, the acid value obtained was 3.945 mgKOH/g. The free fatty acid concentration in the used frying oil was equivalent to 1.973 wt%. Since the free fatty acid content is less than 3 wt%, it implies that the waste frying oil (WFO) could be directly converted into biodiesel via single-step transesterification process. The saponification value was determined to be 183.1 mgKOH/g. This value obtained was lower than that of waste frying oil collected in Malaysia by Tan et al. [44]. Meanwhile, it was approximately equivalent to the one reported by Buasri

Property	Unit	Value
Density at 25°C	g/cm³	0.9147
Viscosity at 40°C	cP	9.36
Acid value	mgKOH/g	3.945
Free fatty acid	wt.%	1.973
Saponification value	mgKOH/g	183.1
Average molecular	g/gmol	939.41

Table 3. Physicochemical properties of waste frying oil (WFO).

et al. [77]. However, the average molecular weight of oil or fat, usually expressed in g/gmol, is a function of acid and saponification values of fat and oil [78]. Average molecular weight of waste frying oil obtained in this study was found to be 939.4 g/gmol, which was comparable to other sources, 942 and 928 g/gmol [75].

4.2. Comparison of physicochemical properties of biodiesel, diesel and their blends

The results of the measured properties of the WFO-derived biodiesel, petrol-diesel and bio-diesel-diesel blends are presented in **Table 4**.

The physicochemical features of different biodiesel-fossil diesel mixtures as contained in **Table 4** indicate that the specific gravities/densities of those blends and pure biodiesel vary in the range of 0.825–0.883. The densities of B20 and B80 samples showed conformance with the ASTM standard (0.86–090) while B50 did not. However, the kinematic viscosities as measured for the blends of biodiesel and conventional diesel samples on comparison with the ASTM standard for biodiesel meet the requirements as they fall within the range (1.9–6.0 mm²/s); this observation is attributed to homogenized mixture, which might have resulted from proper mixing of the two fuels. Meanwhile, these values as seen in **Table 4** are higher than that of fossil diesel (2.03 mm²/s), indicating that biodiesel has large molecular mass [79]. However, the viscosity of pure biodiesel (3.76 mm²/s) was larger compared to those of three blends. This indicates that blending leads to reduction in viscosity. Hence, a complete combustion and reduction in emission of greenhouse gases are possible [80].

Energy content (calorific value) is a property that determines the fuel combustion quality. As can be seen in **Table 4**, biodiesel has lower energy content as compared to that of petrol-diesel. The main reason for this behaviour is due to the fact that biodiesel contains 11% oxygen by

Parameter	Unit	B0	B20	B50	B80	B100
Specific gravity	—	0.809	0.866	0.825	0.872	0.883
Kinematic viscosity	mm²/s	2.03	3.31	2.27	2.23	3.76
Lower heating value (LHV)	MJ/kg	48.23	40.33	47.18	39.33	37.49
Flash point	°C	84	130	79	124	162

Table 4. Physicochemical properties of WFO-derived biodiesel, petrol-diesel and their blends.

Fuel type	CO (ppm)	CO_2 (ppm)	O_2 (ppm)
B0	2300	9500	98,750
B20	1050	14,000	185,250
B50	1055	19,000	173,450
B80	1450	22,500	173,500

Table 5. Average values of exhaust emissions for every biodiesel blend.

weight [81, 82]. More so, it was revealed that B50 has the highest energy content (caloric value) compared among the three blended fuels, followed by B20. However, the lower heating values of those dual fuels (B20, B50 and B80) are higher than that of pure biodiesel but lower than that of petrol-diesel, thus indicating high specific fuel consumption. The same findings were also reported by Yoo and Lee [83] in the prediction models and LHV effect on the CI engine performance when fuelled with biodiesel blends.

The measure of flammability of the fuel is referred to as flash point [7]. The measured values of flash point obtained for the three blended fuels indicate a significant drop when compared to that of B100. This indicates an improvement in fuel qualities. As for diesel fuel (B0), it could be noticed that the results of the specific gravity, the kinematic viscosity, flash point and the lower heating value shown in **Table 4** are very close to the DIN EN 590 standards of diesel and to the experimental results reported by Chopade et al. [84].

4.3. Performance and emission characteristics

In this aspect, the gas exhaust emissions are compared for different biodiesel blends, that is, B20, B50 and B80 and pure diesel (B0) at the engine speed of 2000 rpm. The gas emissions measured include carbon monoxide (CO) and carbon dioxide (CO_2). The gas analyzer used in this study could only measure those aforementioned gases. The mean values of the resulted emissions for each of the fuel are presented in **Table 5**.

As shown in **Table 5**, it was noticed that the average value of CO_2 rises as the volume of biodiesel increases in the mixture of biodiesel and conventional petrol-diesel. This trend may be due to the presence of oxygen in biodiesel [85].

4.3.1. Comparison of carbon monoxide (CO) emission

Higher percentage of CO was emitted when the diesel engine was fuelled with B80 fuel, followed by B50 which released 1055 ppm, and least CO was emitted from diesel engine when it was fuelled with B20 blended fuel as can be seen in **Table 5**. Although, the amount of CO emitted for every fuel biodiesel blend was found to be very small. There was no much difference among the three blended fuels. A similar observation was also recorded by Xue [86], who observed that the blends of biodiesel lowered CO emissions. Moreover, biodiesel, being an oxygenated fuel, enhances combustion and leads to reduction in CO emission. It is noticed that the three blended fuels have lower values of CO as compared to other gases emitted. This indicates that the combustion was almost completely done. However, the reported increase of CO emission with the use of pure diesel (B0) is due to the absence of oxygen in the fuel, thus leading to incomplete combustion [83].

4.3.2. Comparison of carbon dioxide (CO_2) emission

As shown in Table 5, it was noticed that B80 fuel emitted largest concentration of CO_2, followed by B50 fuel, and this indicates that CO_2 emission increases as the biodiesel content increases in the biodiesel-diesel blend. As reported in the literature, biodiesel provides a means of reusing

CO_2, so there is no net increase in global warming [74]. Moreover, it is a known fact that complete combustion inside the combustion chamber of diesel engine promotes CO_2 [7]. Besides, it has been reported by many researchers that the presence of O_2 in biodiesel enhances better combustion [87, 88], which helps to convert CO and CO_2 and, therefore, increases CO_2 emission rate. This is attributed to why B80 fuel released large amount of CO_2 as compared to other blended fuels.

It is quite evident from **Table 5** that the CO_2 emission of biodiesel-fossil fuel mixture is higher than that of pure diesel (B0) at all blends and a maximum increase is noted when compared to each of the blend. According to Datta and Mandal [7], the emission of CO_2 from ignition engine should increase with biodiesel addition to diesel fuel because of improved combustion due to the presence of oxygen in the molecular structure of biodiesel.

5. Conclusions

The major achievements in this research work are the production of biodiesel from WFO via heterogeneous catalyzed transesterification process and the performance of ignition engine fuelled with WFO-derived biodiesel-diesel blend. Various analyses carried out on the WFO-derived biodiesel confirmed that it is of good quality and also meet the ASTM standard. Moreover, the performance and emission evaluation conducted on existing diesel engine fuelled with biodiesel-diesel blend showed that the blend containing 20% biodiesel content (B20) emitted least CO and CO_2, thus suggesting better dual fuel combination.

Acknowledgements

This research was part of my PhD dissertation and was supervised by my co-authors. The authors appreciated the efforts of Prof. A.I. Igbafe (Head, Chemical & Petroleum Engineering, ABUAD, Ado-Ekiti) and Engr. Dr. A. Muktar (Head, Chemical Engineering, FUT, Minna). Ekiti State Ministry of Environment, Ado-Ekiti, Nigeria, is also appreciated.

Author details

Adeyinka Sikiru Yusuff[1]*, Olalekan David Adeniyi[2], Moses Aderemi Olutoye[2] and Uduak George Akpan[2]

*Address all correspondence to: yusuffas@abuad.edu.ng

1 Department of Chemical and Petroleum Engineering, Collage of Engineering, Afe Babalola University, Ado-Ekiti, Nigeria

2 Department of Chemical Engineering, School of Engineering and Engineering Technology, Federal University of Technology, Minna, Nigeria

References

[1] Knothe G, Gerpen JV, Krahl J. The Biodiesel Handbook. Champaign, Illinois: American Oil Chemists' Society Press; 2005

[2] Ma F, Hanna MA. Biodiesel production: A review. Bioresource Technology. 1999;**70**:1-15

[3] Canakci M, Sanli H. Biodiesel production from various feedstocks and their effects on the fuel properties. Journal of Industrial Microbiology and Biotechnology. 2008;**35**(5):431-441

[4] Abdoulmoumine N. Sulphated and hydroxide supported on zirconium oxide catalyst for biodiesel production [published M.Sc thesis]. USA: Faculty of the Virginia Polytechnic Institute and State University; 2010

[5] Knothe GH. The potential contribution of biodiesel with improved properties to an alternative energy mix [abstract]. In: 2nd International Symposium, Kyoto University, Global Centers of Excellence Program; 2010

[6] Refaat AA. Biodiesel production using solid metal oxide catalysts. International Journal of Environmental Science and Technology. 2011;**8**(1):203-221

[7] Datta A, Mandal BK. A comprehensive review of biodiesel as an alternative fuel for compression ignition engine. Renewable and Sustainable Energy Review. 2016;**57**:799-821

[8] Ejaz MS, Younis J. A review of biodiesel as vehicular fuel. Renewable and Sustainable Energy Reviews. 2008;**12**(9):2484-2494

[9] Westberg E. Qualitative and quantitative analysis of biodiesel deposits formed on a hot metal surface [published M.Sc thesis]. Linkoping University: Department of Physics, Chemistry and Biology, Institute of Technology

[10] BS EN 590:2009. Automotive fuels. diesel. Requirements and Test Methods. London, Great Britain: BSI-British Standard Institution; 2009

[11] Sivasamy A, Cheah KY, Fornasiero P, Kemausuor F, Zinoviev S, Miertus S. Catalytic applications in the production of biodiesel from vegetable oils. ChemSusChem. 2009; **2**:278-300

[12] Yee KF, Lee KT. Palm oil as feedstock for biodiesel production via heterogeneous transesterification: Optimization study. In: International Conference on Environment (ICENV). 2008. pp. 1-5

[13] Arponchai SC, Luengnaruemitchai A, Samai JI. Biodiesel production from palm oil using heterogeneous base catalyst. International Journal of Chemical and Biological Engineering. 2012;**6**:230-235

[14] Ibrahim H. Arrangement in heterogeneous catalysis of triglycerides for biodiesel production. International Journal of Engineering and Computer Science. 2013;**2**(5):1426-1433

[15] Cao W, Han H, Zhang J. Preparation of biodiesel from soybean oil using supercritical methanol and co-solvent. Fuel. 2005;**84**:347-351

[16] Noureddini H, Gao X, Phikana RS. Immobilized *Pseudomonas cepacia lipase* for biodiesel fuel production from soybean oil. Bioresource Technology. 2005;**96**:769-777

[17] Yin J, Xiao M, Song J. Biodiesel from soybean oil in supercritical methanol with co-solvent. Energy Conversion and Management. 2008;**49**:908-912

[18] Kouzu M, Kasuno T, Tajika M, Sugimoto Y, Yamanaka S, Hidaka J. Calcium oxide as a base catalyst for transesterification of soybean oil and its application to biodiesel production. Fuel. 2008;**87**:2798-2806

[19] Gelbard G, Bres O, Vargas RM, Vielfaure F, Schuchardt UF. Nuclear magnetic resource determination of the yield of the transesterification of rapeseed oil with methanol. Journal of the American oil chemists' Society. 1995;**72**(10):1239-1247

[20] Kusdiana D, Saka S. Effects of water on biodiesel fuel production by supercritical methanol treatment. Bioresource Technology. 2004;**91**:289

[21] Georgogianni K, Katsoulids A, Pomonis P, Kontominas M. Transesterification of rapeseed oil for the production of biodiesel using homogeneous and heterogeneous catalysis. Fuel Processing Technology. 2009a;**90**(7-8):1016-1022

[22] Jazie AA, Pramanik H, Sinha AS. Eggshell as eco-friendly catalyst for transesterification of rapeseed oil: Optimization of biodiesel production. International Journal of Sustainable Development and Green Economics. 2013;**2**:27-32

[23] Jitputti J, Kitiyanan B, Bunyakiat K, Rangsunvigit P, Jakul PJ. Transesterification of palm kernel oil and coconut oil by difference catalysts. In: The Joint International Conference on Sustainable Energy and Environment (SEE); 1-4; Hua Hin, Thailand. 2004

[24] Demirbras A. Biodiesel fuels from vegetable oils via catalytic and non-catalytic supercritical alcohol transesterifications and other methods: A survey. Energy Conversion and Management. 2003;**44**(2):93-109

[25] Granados ML, Poves MDZ, Alonso DM, Mariscal R, Galisteo FC, Moreno-Tost R, et al. Biodiesel from sunflower oil by using activated calcium oxide. Applied Catalysis B: Environmental. 2007;**73**(3-4):317-326

[26] Arzamendi G, Arguinarena E, Camp I, Zabala IS, Gandia LM. Alkaline and alkaline-earth metals compounds as catalysts for the methanolysis of sunflower oil. Catalysis Today. 2008;**133**(135):305-313

[27] Verziu M, Cojocaru B, Hu J, Richards R, Ciuculescu C, Filip P, et al. Sunflower and rapeseed oil transesterification to biodiesel over different nanocrystalline MgO catalysts. Green Chemistry. 2008;**10**(4):373-381

[28] Dizge N, Aydiner C, Imer DY, Bayramoglu M, Tanriseven A, Keskinler B. Biodiesel production from sunflower, soybean and waste cooking oils by transetserification using lipase immobilized onto a novel microporous polymer. Bioresource Technology. 2009;**100**:1983-1991

[29] Vujicic DJ, Comic D, Zarubica A, Micic R, Boskovic G. Kinetics of biodiesel synthesis from sunflower oil over CaO heterogeneous catalyst. Fuel. 2010;**89**(8):2054-2061

[30] Sun H, Ding Y, Duan J, Zhang Q, Wang Z, Lou H, et al. Transesterification of sunflower oil to biodiesel on ZrO_2 supported La_2O_3 catalyst. Bioresource Technology. 2010; **101**(3):953-958

[31] Ugheoke BI, Patrick DO, Kefas HM, Onche EO. Determination of optimal catalyst concentration for maximum biodiesel yield from tigernut (*Cyperus esculentus*) oil. Leonardo Journal of Sciences. 2007;**6**(10):131-136

[32] Ofoefule AU. Biofuels potentials of some biomass feedstock for bioethanol and biogas [published PhD thesis]. Nsukka, Nigeria: Department of Chemistry, University of Nigeria; 2012

[33] Chen H, Peng B, Wang D, Wang J. Biodiesel production by the transesterification of cottonseed oil by solid acid catalysts. Frontiers of Chemical Engineering in China. 2007; **1**(1):11-15

[34] Kumar V, Kant P. Biodiesel production from sorghum oil by transesterification using zinc oxide as catalyst. Petroleum and Coal. 2014;**56**(1):35-40

[35] Chisti Y. Biodiesel from microalgae. Biotechnology Advances. 2007;**25**:294-306

[36] Armenta RE, Vinatoni M, Burja AM, Kralovec JA, Barrow CJ. Journal of American Oil Chemists Society. 2007;**84**:1045-1054

[37] Mether LC, Dharmagadda VSS, Naik SN. Optimization of alkali-catalyzed transesterification of pongamia pinnata oil for production of biodiesel. Bioresource Technology. 2006;**97**:1392-1397

[38] Kumar D, Kumar G, Poonam G, Singh CP. Ultrasonic-assisted transesterification of *Jatropha curcus* oil using solid catalyst, Na/SiO_2. Ultrasonic Sonochemistry. 2010;**17**(5):839-844

[39] Hawash S, Diwani GEI, Abdel Kader E. Optimization from Jatropha oil by heterogeneous based catalyzed transesterification. International Journal of Engineering and Technology. 2011;**3**(6):1-6

[40] Highina BK, Bugaje IM, Ngala GM. Performance evaluation of continuous oscillatory baffled reactor arrangement production of biodiesel from Jatropha oil using heterogeneous catalyst. World Journal of Renewable Energy and Engineering. 2014;**1**(1):1-7

[41] Helwani Z, Othman MR, Aziz N, Kim J, Fernando WJN. Solid heterogeneous catalysts for transesterification of triglycerides with methanol. Applied Catalysis A: General. 2009;**363**:1-10

[42] Zhang S, Zu YG, Fu YJ, Luo M, Zhang DY, Effert T. Rapid microwave-assisted transesterification of yellow horn corn oil to biodiesel using a heteropolyacid solid catalyst. Bioresource Technology. 2010;**101**(3):931-936

[43] Wen L, Wang Y, Lu D, Hu S, Han H. Preparation of KF/CaO nanocatalyst and its application in biodiesel production from Chinese tallow seed oil. Fuel. 2010;**89**(9):2267-2271

[44] Tan YH, Abdullah MO, Hipolito CN, Taufiq-Yap YH. Waste ostrich and chicken-eggshells as heterogeneous base catalyst for biodiesel production from used cooking oil: Catalyst characterization and biodiesel yield performance. Applied Energy. 2015;**2**:1-13

[45] Lam MK, Lee KT, Mohamed AR. Homogeneous, heterogeneous and enzymatic catalysis for transesterification of high free fatty acid oil (waste cooking oil) to biodiesel: A review. Biotechnology Advances. 2010;**28**(4):500-518

[46] Zhang Y, Dube MA, McLean DD, Kates M. Biodiesel production from waste cooking oil: Process technological assessment. Bioresource Technology. 2003;**89**(1):1-16

[47] Fabbri D, Bevoni V, Notari M, Rivetti F. Properties of a potential biofuel obtained from soybean oil by transmethylation with dimethyl carbonate. Fuel. 2007;**86**(5/6):690-697

[48] Anitha A. Transesterification of used cooking oils catalyzed by CSTPA/SBA15 catalyst system in biodiesel production. International Journal of Engineering and Technology. 2012;**4**(1):34-37

[49] Marchetti JM, Miguel VU, Errazu AF. Possible methods for biodiesel production. Renewable and Sustainable Energy Reviews. 2007;**11**:1300-1311

[50] Banani R, Youssef S, Bezzarga M, Abderrabba M. Waste frying oil with high levels of free fatty acids as one of the prominent sources of biodiesel production. Journal of Material and Environmental Sciences. 2015;**6**(4):1178-1185

[51] Talebian-Kiakalaieh A, Amin NAS, Mazaheri H. A review on novel processes of biodiesel production from waste cooking oil. Applied Energy. 2013;**104**:683-710

[52] Chhetri AB, Watts KC, Islam MR. Waste cooking oil as an alternate feedstock for biodiesel production. Energies. 2008;**1**:3-18

[53] Radich A. Biodiesel Performance, Costs, and Use. US Energy Information Administration. 2006. Available from: http://www.eia.doe.gov/oiaf/analysispaper/biodiesel/idexhtml

[54] Supple B, Holward-Hildige R, Gonzalez-Gomez E, Leashy EJJ. The effect of stream treating waste cooking oil on the yield of methyl ester. Journal of the American Oil Chemists' Society. 2002;**79**(2):175-178

[55] Adepoju TF, Olawale O. Acid catalyzed esterification of waste cooking oil with high FFA for biodiesel production. Chemical and Process Engineering Research. 2014;**21**:80-85

[56] Canakci M, Van Gerpen J. Biodiesel production from oils and fats with high free fatty acids. Transaction of the American Society of Agricultural Engineering. 1999;**44**(6):1429-1436

[57] Charoenchaitrakool M, Thienmethangkoon J. Statistical optimization for biodiesel production from waste frying oil through two-step catalyzed process. Fuel Processing Technology. 2011;**92**:112-118

[58] Lam MK, Lee KT, Mohamed AR. Homogeneous, heterogeneous and enzymatic catalysis for transesterification of high free fatty acid oil (waste cooking oil) to biodiesel: A review. Biotechnology Advances. 2010;**28**(4):500-518

[59] Veljkovic VB, Lakicevic SH, Stamenkovic OS, Todorovi ZB, Lazic ML. Biodiesel production from Tobacco (*Nicotiana tabacum* L.) seed oil with a high content of free fatty acid. Fuel. 2006;**85**:2675

[60] Hanny JB, Shizuko H. Biodiesel production from crude *Jatropha curcas* L. seed oil with a high content of free fatty acids. Bioresource Technology. 2008;**49**:1716-1721

[61] Jayasinghe TK, Sungnornpatansakul P, Yoshikaw K. Enhancement of pretreatment of process for biodiesel production from Jatropha oil having high content of free fatty acids. International Journal of Energy Engineering. 2014;**4**(3):118-126

[62] Kulkarni MG, Dalai AK. Waste cooking oil—An economical source for biodiesel: A review. Industrial and Engineering Chemistry Research. 2006;**45**(9):2901-2913

[63] Sani YM, Daud WMAW, Abdul Aziz AR. Biodiesel feedstock and production technologies, successes, challenges and prospects. Intech. 2013;**4**:77-101

[64] Isayama Y, Saka S. Biodiesel production by supercritical process with crude biomethanol prepared by wood gasification. Bioresource Technology. 2008;**99**:4775

[65] Shuit SH, Yit TO, Keat TL, Bhatia S, Soon HT. Membrane technology as a promising alternative in biodiesel production: A review. Biotechnology Advances. 2012;**30**(6):1364-1380

[66] Freedman B, Butterfield RO, Pryde EH. Transesterification kinetics of soybean oil. Journal of the American Oil Chemists' Society. 1986;**63**(10):1375-1380

[67] Abdulla R, Ravindra P. Immobilized *Burkholderia Cepacia* lipase for biodiesel production from crude *Jatropha Curcas* oil. Biomass and Bioenergy. 2013;**56**:8-13

[68] Borges ME, Alvarez-Galvan MC, Brito A. High performance heterogeneous catalyst for biodiesel production from vegetal and waste oil at low temperature. Applied Catalysis B, Environmental. 2011;**102**(1):310-315

[69] Jaya N, Ethirajulu K. Kinetic modeling of transesterification reaction for biodiesel production using heterogeneous catalyst. International Journal of Engineering Science and Technology. 2011;**3**(4):3463-3466

[70] Jitputti J, Kitiyanan B, Rangsunvigit P, Bunyakiat K, Attanatho L, Jenvanitpanjakul P. Transesterification of crude palm kernel oil and crude coconut oil by different solid catalysts. Chemical Engineering Journal. 2006;**116**(1):61-66

[71] Olutoye MA, Hameed BH. A highly active clay-based catalyst for the synthesis of fatty acid methyl ester from waste cooking palm oil. Applied Catalysis A: General. 2013;**450**:57-62

[72] Liu X, He H, Wang Y, Zhu S. Transesterification of soybean oil to biodiesel using SrO as a solid base catalyst. Catalysis Communications. 2007;**8**(7):1107-1111

[73] Tesfa B, Gu F, Mishra R, Ball A. LHV prediction models and LHV effect on the performance of CI engine running with biodiesel blends. Energy Conversion and Management.
2013;**71**:217-226

[74] Elsolh NEM. The manufacture of biodiesel from the used vegetable oil [published mas-
ter thesis]. Egypt: Faculty of Engineering at Kassel and Cairo Universities; 2011

[75] Mahgoulb HA, Salih NA, Mohammad AA. International Journal of Multidisciplinary
and Current Research. 2015;**3**:1-6

[76] Abdullah N, Hassan SH, Mohd Yusoff MR. Biodiesel production based on waste cooking oil (WCO). International Journal of Materials Science and Engineering. 2013;**1**:1-7

[77] Buasri A, Chaikwan T, Loryuenyong V, Rodklum C, Chaikwan T, Kumphan N. Continuous process for biodiesel production in packed bed reactor from waste frying oil
using potassium hydroxide supported on *Jatropha Curcas* fruit shell as solid catalyst.
Applied Science. 2012;**2**:641-653

[78] Zhu H, Wu Z, Chen Y, Zhang P, Duan S, Liu X, et al. Preparation of biodiesel catalyzed by solid super base of calcium oxide and its refining process. Chinese Journal of
Catalysis. 2006;**27**(5):391-396

[79] Jaichandar S, Annamalai K. The status of biodiesel as an alternative fuel for diesel
engine—An overview. Journal of Sustainable Energy and Environment. 2011;**2**:71-75

[80] Mythili R, Venkatachalam P, Subramaniam P, Uma D. Production characterization and
efficiency of biodiesel: A review. International Journal of Energy Research. 2014;**38**(10):
1233-1259

[81] Demirbas A. Biodiesel: A Realistic Fuel Alternative for Diesel Engines. London: Springer-
Verlag London Limited; 2008

[82] Nair P, Singh B, Upadhyay SN, Sharma YC. Synthesis of biodiesel from low FFA waste
frying oil using calcium oxide derived from Mereterix as a heterogeneous catalyst.
Journal of Cleaner Production. 2012;**29**(30):82-90

[83] Yoo SH, Lee CS. Experimental investigation on the combustion and exhaust emission
characteristics of biogas-biofuel dual-fuel combustion in a CI engine. Fuel Processing
Technology. 2011;**92**:992-1000

[84] Chopade SG, Kulkarni KS, Kulkarni AD, Topare NS. Solid heterogeneous catalysts
for production of biodiesel from transesterification of triglycerides with methanol: A
review. Acta Chimica and Pharmaceutica Indica. 2012;**2**(1):8-14

[85] Miraculus GA, Bose N, Edwin RR. Optimization of biofuels blends and compression
ratio of a diesel engine fueled with *Calophyllum inorphyllum* oil methyl ester. Arabian
Journal of Science and Engineering. 2016;**4**(5):1723-1733

[86] Xue J. Combustion characteristics, engine performances and emissions of waste edible oil
biodiesel in diesel engine. Renewable and Sustainable Energy Reviews. 2013;**23**:350-365

[87] Godiganur S, Murthy CHS, Reddy RP. GBTA 5.9 G2-1 Cummins engine performance and emission tests using methyl ester mahua (*Madhuca indica*) oil/diesel blends. Renewable Energy. 2009;**344**:2172-2177

[88] Nabi MN, Hoque SMN, Akhtar MS. Karanja (*Pongamia pinnata*) biodiesel production in Bangladesh, characterization of Karanja biodiesel and its effect on diesel emissions. Fuel Processing Technology. 2009;**90**:1080-1086

High-Octane Gasoline Production from Catalytic Naphtha Reforming

Cheng Seong Khor

Additional information is available at the end of the chapter

http://dx.doi.org/10.5772/intechopen.80866

Abstract

The global drive for environmental sustainability necessitates continuous adjustment, optimization, and improvement in petroleum refining processes to generate energy and products including automotive fuels such as gasoline. At the same time, refiners need to maximize their asset utilization to maintain competitiveness in the business setting. This chapter presents a process advisory and monitoring application to optimize a catalytic naphtha reforming operation to produce high octane gasoline feedstock. A mathematical model is developed for the process to produce hydrocarbons with high anti-knock ratings. The proposed methodology involves formulating a nonlinear programming optimization model to perform data reconciliation. The model objective minimizes the deviations (or errors) between the measured values and the model-reconciled values to reflect the accuracy and reliability of the measurements. The overall procedure is carried out subject to various real-world operation constraints to ensure sustainable processing of the required products, which include hydrogen gas and aromatics. We present a case study to illustrate an implementation of the resulting model in an online environment to improve process operation at an actual refinery in Canada. The computational results show enhanced product quality of a reformate stream with high octane number and increased yields.

Keywords: catalytic naphtha reforming, petroleum refinery process operation, data reconciliation, process monitoring, mathematical modeling, nonlinear optimization

1. Introduction

1.1. Catalytic reforming process

Catalytic reforming is an important process in the petroleum refining industry. It is developed originally to produce components of automotive fuels specifically gasoline, which meet engine requirement for high antiknock quality. The process objective is to convert petroleum naphtha fractions to high rates of aromatic hydrocarbons as selectively as possible since the latter have excellent antiknock ratings. Naphtha fractions are liquid hydrocarbon mixtures with 6–12 carbon atoms and having boiling points in the range of 320–470 K [1, 2]. Reforming also serves two other main purposes in a refinery: it is the main hydrogen producer for use within a refinery or outside it; it also provides feedstock (mainly consists of benzene, toluene, and xylene) for the subsequent downstream petrochemical production processes [3]. A survey of the recent progress on the reforming process focusing on the reactor modeling is available in Refs. [4].

A commercial reformer unit consists of a reactor section, a recycle gas compression section, and a fractionation section. The reactor section consists of a feed system, a few heaters or furnaces, a series of reactors, and a flash separator. A portion of the flashed hydrogen gas is recycled and mixed with the feed and then increased to the reaction temperature by a heat exchanger combining the feed and reactor effluent followed by the first heater. We send the flashed liquid to the fractionation section comprising a distillation column that acts as product stabilizer. The distillate stream strips light gases from the flashed liquid that produce liquefied petroleum gas (LPG) and off-gas as fuel. The main product is the bottoms stream called reformate that is a feedstock for gasoline blending. We use a few (typically three to five) heater-and-reactor pairs to maintain the reaction temperature within a range of 400–500°C (700–800 K) and at pressures of 10–35 atm using catalysts mainly to accelerate the reaction [3, 6]. **Figure 1** displays a general process flow of a commercial reformer unit.

1.2. Data reconciliation method

Data reconciliation involves estimating process variables by comparing their values from process measurements and process models. The models typically comprise material and energy balances or conservation relations. Data reconciliation involves adjusting or correcting errors in measurement values to be consistent with the mass balances. The procedure is useful in the process and its affiliated industries to conduct monitoring and modeling for control, simulation, and optimization as well as to perform instrument maintenance and equipment analysis [5–9].

Progress has been recorded in the literature on data reconciliation techniques that include calculating measurement errors to estimate the model variable values using linear process models [7], classifying the estimated variables [8], and categorizing the measurement values as redundant or nonredundant [9]. In this work, we adopt the data reconciliation procedure that makes use of information from redundant measurement and conservation laws to correct measurement values by converting them into model values based on reliable, accurate, and thermodynamically valid process knowledge incorporated in a nonlinear steady-state model.

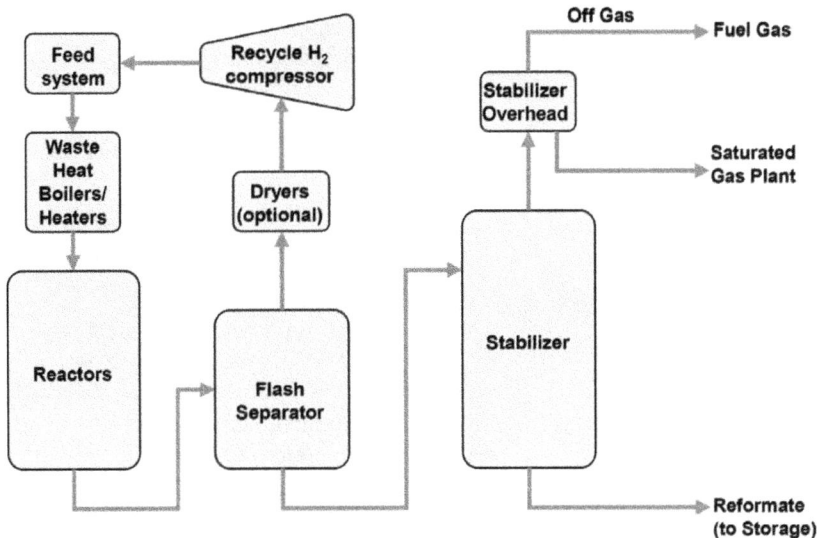

Figure 1. General process flow of a commercial reformer unit.

2. Problem statement

We have the following data for a catalytic reforming process: a set of process units; a set of measurements i with vector of known data on their raw scan values y, model values x, standard deviation σ_i, and weights w_i; and a set of tuning parameters p with known nominal values n_p, model values t_p, and scaling factors s_p. The model constraints for the process are the mass balances, energy balances, and equilibrium relationships.

In this work, we want to optimize the operation of a catalytic reforming process by minimizing the deviations of values for selected measurements and tuning parameters from their model-computed values (i.e., differences computed between the x and y vectors and the n_p and t_p vectors, respectively) subject to process constraints to obtain reconciled process values (as given by the model values). To achieve this aim, we can adjust a set of reconciled variables such as skew factors on the true boiling point of crude oil fractions, product flow rate bias, Murphree efficiencies of the distillation columns, and reaction kinetic parameters [10–12].

3. Model formulation

This section describes the approach used to develop a reforming process monitoring model using the commercial advanced chemical process optimizer platform called SimSci ROMeo [13]. The model developed involves two technologies: cyclic and semi-regenerative reforming with the overall scope covered by the model shown in **Figure 2**. **Figure 3** presents a detailed schematic of the model for the cyclic reforming operation.

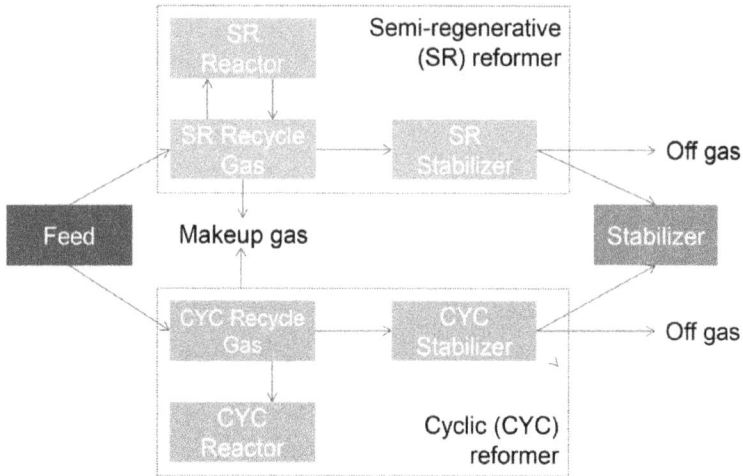

Figure 2. Model scope comprising the semi-regenerative and cyclic reforming units.

The optimality of a catalytic reforming process operation is measured based on the yields of the main product of reformate (a gasoline blending feedstock) and the side products of petrochemicals comprising benzene, toluene, and xylene (BTX). We use material-balanced data of the model-computed values and not raw scan values of the measurements to evaluate such performance measures. Hence, we perform data reconciliation on the measurement values, which are obtained from field instruments measuring the actual process operation.

3.1. Data reconciliation model

We formulate the data reconciliation procedure as a least squares minimization version of an optimization problem [14]. The objective function as given in Eq. (1) minimizes the sum of squares of the weighted differences between model values and raw scan values for selected measurements and tuning parameters:

$$\text{minimize} \sum_i w_i \left(\frac{\delta_i}{\sigma_i}\right)^2 + w_p \left(\frac{t_p - n_p}{s_p}\right)^2 \tag{1}$$

where w_i is the weight matrix for each measurement i (either 1 or 0 otherwise for those not considered) and δ_i is the offset as given by difference between model value and raw scan value.

The differences or deviations are also called errors, offsets, or biases of the instruments considered. The offset value is weighted or multiplied with a weight factor (i.e., its value is $w_i = 1$) if we decide to reconcile the process value of the associated measurement or tuning parameter. Such a decision is made based on a measurement's reliability, that is, if it is measured value is accurate to be compared against a model-reconciled value; otherwise, the instrument needs to be calibrated. The model constraints mainly consist of the total mass and component balances, energy balances, and equilibrium relationships as well as appropriate bounds on the variables [7, 8].

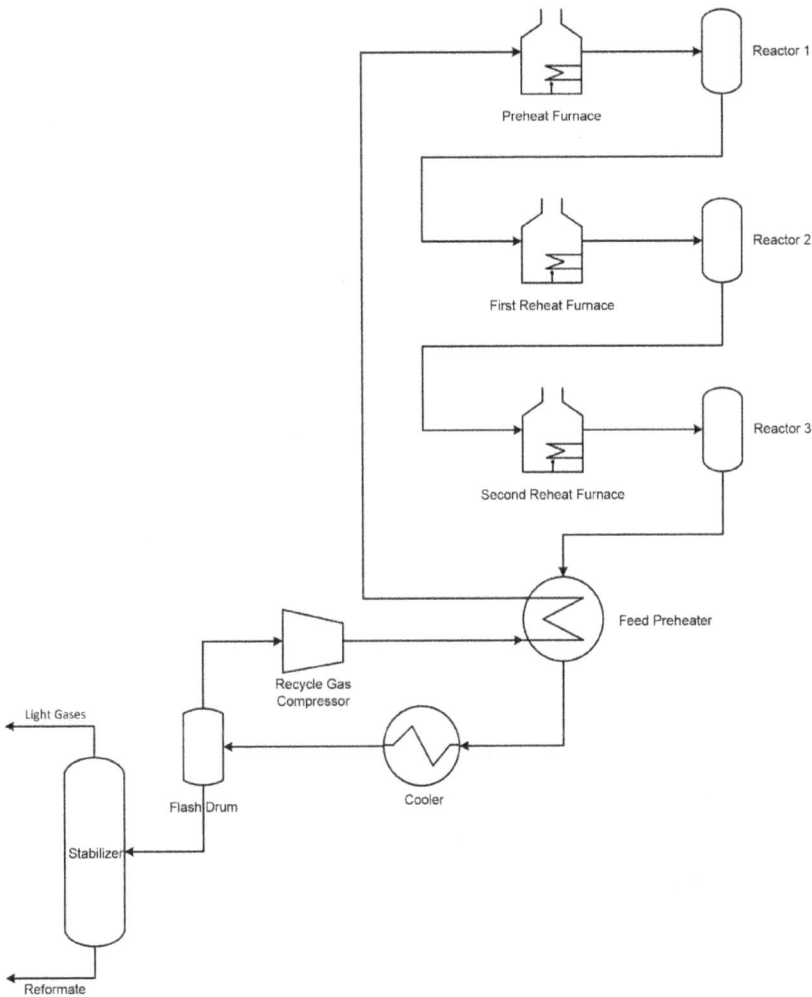

Figure 3. Schematic representation of the cyclic reforming operation model representation.

3.2. Feed characterization

We perform tuning on the reformer feed composition to match the process data as close as possible by adjusting the factors representing the component mole fractions [10]. It is not necessary to adjust for all the components in the reconciliation procedure especially if regular laboratory data on feed composition is not available in terms of its PONA (paraffin, olefin, naphthene, and aromatic) content analysis. We can reduce the number of components to be tuned for the feed slate by equating the mole fraction factors for components with the same number of carbon atoms on the basis that these components have similar molecular weights and densities. This strategy can result in reducing about one-third of the parameters involved in the tuning procedure.

In the absence of composition data, we consider tightening the reformate yield specification by minimizing its offset and reducing its standard deviation value. The reformate yield measured value is given by the ratio between reformate mass flow rates to that of the feed; therefore, no weight is specified on reformate yield because the actual measured values are the product and feed flows.

In a component adjuster model of the SimSci ROMeo software, three choices of methods are available for handling each component, that is, flow, component, and dependent. Typically, only one component is chosen as a dependent component; all others should be chosen as flow or component. There is no definite guideline on specifying the use of a flow or component method for a component. It is best to use the most abundant component which is the dependent component.

Depending on the situation, we can impose bounds on the adjustment factor to get a feasible solution. If the lower bound of the adjustment factor for a component is active, setting its lower bound to a value of −0.5 results in reducing the component mole fraction by 50% (e.g., from 0.05 to 0.025). Correspondingly, a lower bound of −1 reduces the component mole fraction to zero, that is, the component is removed from the resultant outlet stream of the component adjuster. This effect is undesirable; we do not want a component to be removed completely. Therefore, a suggestion is to use a lower bound other than −1 (e.g., −0.8 or −0.75). Similarly, for a component with an active upper bound, an upper bound of 0.5 increases its mole fraction by 50% (e.g., from 0.05 to 0.075). Therefore, we avoid upper bounding of the adjustment factor on the mole fraction of a component as 1 because it results in a doubling effect.

It is extensive and not necessary to tune the compositions of all components in the lumped feed slate. In particular, there is no need to adjust those of the light components of C5s and lower (i.e., C3s–C5s) because their compositions are very low (or almost zero). A typical bound for the tuning parameters of a feed component is [−0.75, 0.75]; a practical bound depends on the actual data.

3.3. Reactor representation

The application uses a proprietary nonlinear surrogate model to represent a reactor bed that incorporates its catalyst- and kinetic-related information. We use a flash drum operating under adiabatic condition to model the pressure of the reactor bed inlet stream and a similar configuration to model the outlet stream as shown in **Figure 4**. For each reactor, we include a set of

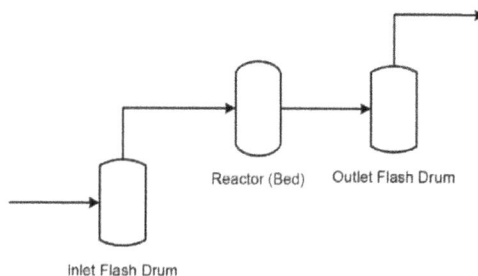

Reactor (Bed)   Outlet Flash Drum

Inlet Flash Drum

Figure 4. Schematic representation of a reactor.

equations to equate each of the kinetic parameters (e.g., catalyst activity) to a common variable, thereby facilitating the reaction kinetics tuning to match the plant values. Since each reactor is identical, we first develop an individual reactor with its associated auxiliary units (such as the two flash drums at its inlet and outlet) and then duplicate its representation. This approach can be done by way of using a block diagram (or other similar features) that is available, which can help to improve the layout of the graphical user interface besides facilitating with a failed solution or modeling problem.

3.4. Reactor pressure balance

There are a number of ways to represent a reactor pressure balance for data reconciliation purpose depending on the reliability of measurements available: (1) distribute the total pressure drop across the reactor bed equally to the pressure drop of the inlet flash drum and that of the outlet flash drum; (2) honor the inlet and outlet pressure measurements if they are reliable; and (3) relate the pressure drops for the other reactors to that of a reactor with known reliable pressure measurement.

3.5. Reaction kinetic tuning

A general approach in reforming applications is to maintain most of the built-in reaction kinetic tuning parameters equal for each of the reactors by using some form of mathematical relation facility (such as a customization unit feature available in ROMeo). We use the first mathematical relation with a local variable on each kinetic parameter that has an associated tuning parameter to adjust it. Then, for each reactor, we use the second relation to equate a kinetic parameter to its corresponding variable in the first relation. Such a setup gives the flexibility of turning a reactor on or off when addressing convergence issues while keeping the kinetic tuning parameters equal for all reactors that are turned on. We put suitable bounds on the tuning parameters to keep the solved parameter values within reasonable range. The tuning parameters are divided into two sets: for catalyst and for catalyst coke. The catalyst tuning parameters are base-2 logarithm ($\log_2$) multipliers that are bounded.

3.6. Reactor switch in cyclic reformer

The application has logic to represent the reactor switch for a cyclic reformer. The switch operation involves a swing reactor besides the on-oil reactors to replace the reactor in regeneration. Thus, the catalyst can be regenerated without shutting the unit down. Note that such logic is not required for a semi-regenerative reformer in which all the reactors are taken off-stream or out of service for in situ regeneration of the entire catalyst inventory.

The mentioned logic to model the reactor switching involves the following steps with the explanation aided by **Figure 5**:

- Turns down the following items for the off-oil reactor and turns them on for the on-oil reactors: inlet pressure measurement, reactor custom model unit, inlet and outlet flash drums, customization units on reaction kinetic tuning, customization unit on reactor pressure balance, delta temperature measurement model, and its associated customization unit

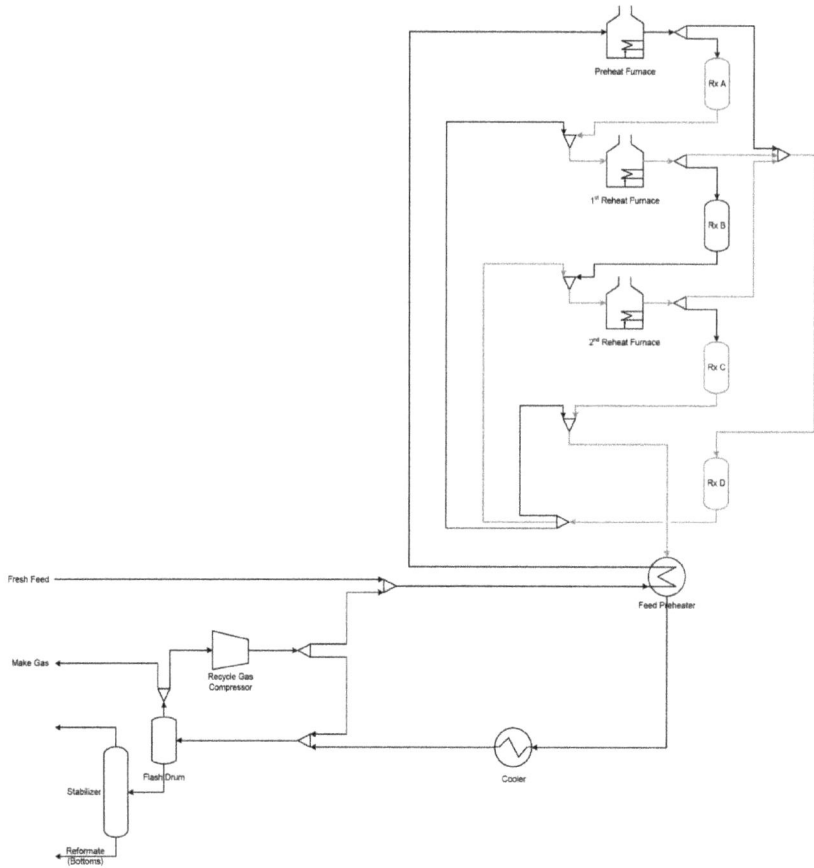

Figure 5. Reactor switching in cyclic reformer.

- Sets to zero for the initial and final values of the off-oil reactor bed pressures (and equates these values to that of the inlet flash drum pressures for the on-oil reactors)

- Sets to zero for the split fraction of the off-oil reactor inlet flow rate: sets to 1 for the split fraction of the swing reactor bypass flow rate and vice versa for the on-oil reactors, thereafter generating estimates for these splitters

If the swing reactor is off oil, the logic performs the following steps:

- Turns on the inlet mixer and outlet splitter of the swing reactor

- Initializes the mixer pressure drop to that of the incoming bypass stream, thereafter generating value estimates

- Sets the split fractions for the swing reactor outlet splitter flow rates, thereafter generating value estimates

- Initializes the mixer pressure drop for each on-oil reactor to the incoming stream, thereafter generating value estimates

- Unweight the inlet pressure and temperature difference measurements

The default setting of the logic is to put weights on all the reactor inlet pressure and temperature difference measurement models to account for them in the objective function computation. But when reactor switch happens, the swing reactor inlet pressure measurement may have a large offset that results in a huge objective value; hence, it may merit resolving the model by removing the weight on this tag.

3.7. Measurement models

3.7.1. Absolute error tolerance

The standard deviation parameter is specified to set the absolute error tolerance for a measurement model. Manipulating this tolerance to obtain a desired solution or to represent the importance of a measurement should be avoided; instead, the tolerance represents a measurement's reliability and accuracy.

We use Microsoft Excel to view and analyze the results to help with obtaining a good representation of the measurements; an example of such a result viewer is shown in **Figure 6**. A measurement may still be of good quality but a particular instance of its value may be bad, for example, because data has stopped flowing to the data historian. Specifying the tolerance of a measurement depends on its precision. The weighting of a measurement can handle a bad valued measurement instance by way of excluding it from being considered in the objective function computation.

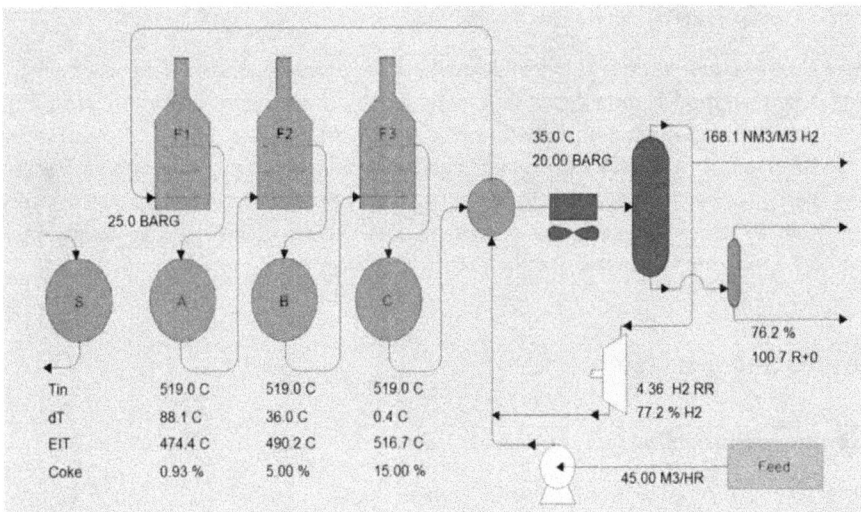

Figure 6. An example of result viewer of the reformer model.

It is recommended to set approximately 1% of the process variable value (PV) for most measurements and 5% of the PV for flow rates. Note that a calculated tag is generally not reconciled, that is, unweighted if the input measurement for which the calculations are based on are available.

3.7.2. Measurement screening logic to handle data quality

Specifying the logic for measurement model screening requires review when handling negative flow rate instances to determine if they are caused by zero flows or bad transmitters. The following gives a list of rules of thumb that can be applied:

- The typical response to bad quality value is to use the last good raw scan value, which applies particularly for a measurement model whose value is a dependent variable. For temperature indicators (thermocouples), we can select to set its objective function contribution to zero.

- If we do not want a measurement value to become negative when its quality turns bad, we can fix its value to zero or a small value by setting its minimum and selecting an out-of-range action. This rule typically applies to a flow rate meter which we do not need to specify a maximum value.

- Set zero objective function contribution for the offset of a dependent valued measurement.

- To respond to bad quality data for an independent valued measurement, we can use a suitable fallback value as the scan value or stop running the online model to find out the cause especially if this happens to a feed flow rate meter.

- To fix controller outputs at certain minimum and maximum values.

3.7.3. Two measurements on one variable

There may be two measurements available on the same process variable. It is acceptable to use two measurement models for a dependent variable, but for an independent variable, doing so results in one measurement model used as a controller and another as an indicator; thus, we may face problems if its scan value becomes bad. If both measurement models are necessary, we can specify a new variable equal to the particular process variable and point the indicator measurement to it, besides specifying a screening criterion to set the objective value contribution to zero when the measurement data quality turns bad.

4. Results and discussion

4.1. Key process variables

The key process variables to match are reformate yield, reformer reactor total endotherm, research octane number (RON) of reformate product, and hydrogen recycle gas purity. The available tuning variables are feed composition and reaction kinetic parameters such as overall catalyst activity, acid site activity, and aromatic selectivity. The main input parameters are catalyst weight and reactor bed coke fraction. **Table 1** summarizes the key variables that the application uses for process monitoring and their typical values.

4.2. Tuning strategies

To tune the model to achieve the desired values and results, we balance between tuning the group of parameters for the reactor kinetics and that for the feed compositions. We can reduce the variation in one group by letting parameters in the other group move—the deliberation as to which to allow for more movement/adjustment depends on the feedback from the site or a process specialist on their relative importance. If the deliberation is not to change the feed composition much while letting the reactor tuning parameters to move more, then use scaling factors of 1 for both parameter groups with appropriate lower and upper bounds for the parameters. (The solved parameters should not be at bounds.)

However, if the composition of the synthesized feed is significantly different from the actual plant condition, even allowing for the feed composition tuning parameters to move over a wide range may not help with convergence. Therefore, it is important to have good starting feed composition, which can be achieved by calculating the feed distillation cut points at the same time at which the samples are taken.

Two other tuning strategies involve using calculated variables on the reformate yields and reactor total endotherms.

Variable	Error	Value (offset in bracket)	
		Cyclic	Semi-regenerative
Reformate yield	±0.1%	81.19 (−0.13)%	82.68 (−0.26)%
Reactor total endotherm	±10°F	198.17 (−0.01)°F	152.07 (−5.20)°F
Research octane number (RON)	±0.1	87.39 (−0.35)	87.53 (−0.63)
Hydrogen recycle purity	±1%	90.02 (−3.23)%	75.60 (0.52)%
Reaction kinetic parameter	Reference		
Catalyst activity	[−100, 100]	1.13	2.30
Metal activity	[0.5, 1.5]	1.01	1.45
Acid activity	[0.5, 1.5]	1.14	1.28
Aromatic selectivity	[0.5, 1.5]	1.11	1.65
Hydrogenolysis activity	[0.5, 1.5]	1.25	1.18
Catalyst coke capacity	[0.5, 1.5]	1.26	1.73
Coke yield penalty	[0.5, 1.5]	1.52	0.78
Coke hydrogen purity penalty	[0.5, 1.5]	1.10	1.60

Table 1. Representative key process variable values to match error tolerance for reformate yields.

4.2.1. Reformate yields

Reformate yield is given by the ratio of the reformate stream flow rate to that of the feed stream (typically on a mass basis). A calculated tag is generally not weighted and thus excluded from the objective function calculation. Instead, the tolerances for the two flow measurements used to calculate the yield (i.e., for the reformate and feed streams) are tightened. However, for the purpose of experimenting to determine ways to improve the model, we put a weight on the reformate yield tag and use a tight tolerance by specifying the standard deviation value to an artificially small number but one in which the model is still solvable in reasonable computational time. The resulting model solution gives information on the candidate independent variables whose bounds can be reasonably relaxed, measurement models whose tolerances (i.e., standard deviations) can be adjusted or relaxed, as well as possibly faulty or problematic instrumentation (due to malfunction or calibration issues). Note that this practice highlights a difference between using a data reconciliation application and using a predictive optimization tool in which the latter attempts to meet the reformate yields and other variables tightly.

4.2.2. Reactor total endotherms

Reactor total endotherms are the sum of the temperature differences (i.e., delta in temperatures) of each of the reforming reactors. This variable represents the temperature change and hence the reaction that takes place in a reactor, which can be the desirable endothermic reactions of dehydrogenation of naphthene to aromatic compounds and/or dehydrocyclization of paraffins to naphthenes or the undesirable (but necessary) exothermic reactions of hydrocracking and/or hydrocyclization. There are two relations available (which hence gives rise to uncertainty) to calculate the temperature difference for a reactor: (1) take the difference between the temperature transmitters at the outlet of a preheater (i.e., upstream furnace) and that at the inlet of a furnace downstream and (2) take the difference between the reactor temperature indicators (i.e., thermocouples at the top skin and bottom skin). It is noted that when facing catalyst issues, the second relation may give a higher value than that theoretically attainable due to heat loss that is actually experienced but not accounted for. Similar to the reformate yield being a calculated variable, it is also generally our practice to unweight (or to turn off) the total endotherm tag because it is an artificial or pseudo-tag which may not physically exist at a site.

5. Concluding remarks

In this work, we discuss a systematic workflow to develop a data reconciliation model for a catalytic reforming process to produce automotive fuels. We present the mathematical model formulation in the form of a nonlinear optimization model developed on a commercial software platform. The model is deployed as an online application to improve process advisory and monitoring at a refinery besides providing a rich data source for a multitude of other associated relevant uses. We show the utility of such an application to contribute toward producing energy in an environmentally sustainable manner through an optimal process operation approach with reduced off-spec fuel products.

Author details

Cheng Seong Khor[1,2]*

*Address all correspondence to: khorchengseong@gmail.com

1 Chemical Engineering Department, Universiti Teknologi PETRONAS, Seri Iskandar, Perak Darul Ridzuan, Malaysia

2 Academy of Sciences Malaysia, Kuala Lumpur, Malaysia

References

[1] Moser M, Sadler CC. Reforming—Industrial. In: Horváth IT, editor. Encyclopedia of Catalysis. New Jersey: Wiley; 2010. DOI: 10.1002/0471227617

[2] Moser MD, Bogdan PL. Catalytic reforming. In: Ertl G, Knözinger H, Schüth F, Weitkamp J, editors. Handbook of Heterogeneous Catalysis. Weinheim, Germany: Wiley-VCH Verlag; 2008

[3] Sinfelt JH. Catalytic reforming of hydrocarbons. In: Anderson JR, Boudart M, editors. Catalysis–Science and Technology. Vol. 1. Berlin, Heidelberg: Springer-Verlag; 1981. pp. 258-298

[4] Rahimpour MR, Jafari M, Iranshahi D. Progress in catalytic naphtha reforming process: A review. Applied Energy. 2013;**109**:79-93

[5] Heyen G, Kalitventzeff B. Process monitoring and data reconciliation. In: Puigjaner L, Heyen G, editors. Computer Aided Process and Product Engineering. Weinheim: Wiley-VCH Verlag; 2006

[6] Amand T, Heyen G, Kalitventzeff B. Plant monitoring and fault detection: Synergy between data reconciliation and principal component analysis. Computers and Chemical Engineering. 2001;**25**:501-507. DOI: 10.1016/S0098-1354(01)00630-5

[7] Wang D, Romagnoli JA. Generalized T distribution and its applications to process data reconciliation and process monitoring. Transactions of the Institute of Measurement and Control. 2005;**27**(5):367-390

[8] Lid T, Skogestad S. Data reconciliation and optimal operation of a catalytic naphtha reformer. Journal of Process Control. 2008;**18**:320-331. DOI: 10.1016/j.jprocont.2007.09.002

[9] Wu SX, Ye Q, Chen C, Gun XS. Research on data reconciliation based on generalized T distribution with historical data. Neurocomputing. 2016 Jan;**175**:808-815

[10] Britt HI, Luecke RH. The estimation of parameters in nonlinear, implicit models. Technometrics. 1973;**15**(2):233-247

[11] Stanley GM, Mah RSH. Observability and redundancy in process data estimation. Chemical Engineering Science. 1981;**36**(2):259-272

[12] Crowe CM. Observability and redundancy of process data for steady state reconciliation. Chemical Engineering Science. 1989;**44**(12):2909-2917

[13] Schneider Electric. ROMeo Process Optimization Solution: Rigorous Online Modeling with Equation-Based Optimization [Internet]. Available from: http://software.schneider-electric.com/pdf/datasheet/romeo-process-optimization. [Accessed: June 05, 2018]

[14] Gill PE, Murray W, Wright MH. Practical Optimization. London, UK: Academic Press; 1981

Implementation of Basic Principles of Econometric Analysis in Petroleum Technology: A Review of the Econometric Evidence

Constantinos Tsanaktsidis and
Konstantinos Spinthiropoulos

Additional information is available at the end of the chapter

http://dx.doi.org/10.5772/intechopen.80510

Abstract

In the present study we give the opportunity to understand the physicochemical parameters of distilling petroleum products applying the basic principles of econometric analysis. The quality of the different fuels is expressed by a series of physical, chemical or other characteristics. The connection between production process and quality of fuel is crucial in the field of petroleum technology. It is remarkable that the used method of the regression analysis perfectly illustrates the relationships between the variables in all applied models. Econometrics is one the best methods to study the variation of the physicochemical properties of the oil. The use of econometrics methods in petroleum chemistry turned out to be useful tool in order to prove that there is indeed strong rates volatility and correlation between physicochemical properties of oils with their mixes. In Petroleum Industry the most common types of Diesel fuels are the biodiesel, biomass to liquid or gas to liquid Diesel. The results of our research can be an important tool for the development of software that can anticipate changes of physicochemical properties of petroleum distillate products, taking into account specific parameters.

Keywords: diesel, JP8, biodiesel, econometric, analysis

1. Introduction

Econometric applications are now accepted in order to have safe conclusions in oil technology as well. The use of such applications can provide the necessary information to deal with problems, particularly in terms of fuel quality. The forecast is based on mathematical equations that take

IntechOpen

into account specific fuel price constraints. In this way, it is possible to check the quality of the fuel during the production process and also the quality at its distribution points.

Econometric analysis is used in order to find out whether there is a relationship or not between particular variables. The relative evaluation realization demands the application of statistic and mathematic methods which are focused on the variables features used by the analysis.

The use of linear regression analysis is one of the most famous econometric methods. By using structure and unstructured data we can evaluate empirical research and taking into account the theory we can come up with safe conclusions [1]. An introductory economics textbook describes econometrics as allowing economists "to sift through mountains of data to extract simple relationships [2]. The first known use of the term "econometrics" was by Polish economist Pawel Ciompa in 1910 [3].

As already mentioned above the basic tool for econometric analysis is the linear regression method. With this model we can create models and come up to safe conclusions. The linear regression method can also be easily applied in the case of oil or pure diesel blends. In modern econometrics, other statistical tools are frequently used, but linear regression is still the most widespread method among all. Especially for the study of the physicochemical properties of the oil, meaning diesel and biodiesel, we are dealing with non-chronological data. Econometrics is one the best methods to study the variation of the physicochemical properties of the oil in order to draw strong and reliable conclusions about the quality of the examined mixture and in most cases to able to preview the values of its physicochemical property of the tested model fuels [4]. Estimating a linear regression on two variables can be visualized as fitting a line through data points representing paired values of the independent and dependent variables [4].

Greene [5] suggests that the environmental and energy issues increase the demand for different fuel types of altered and mixed oils. The market of petroleum products is enlarged as a consequence of the changes in the environment of life that have originated from the improvement of the living conditions, the cities expanding as a result of the massive population movement and the expanded basic needs.

Next to this, the national trade balance and the economic development are influenced by this compulsory demand since Greece imports more crude oil products owing to the expansive requirement. Additionally, the transportation energy use of diesel causes a serious environmental pollution by its gases, leading EU countries to command against them and avoid the bad side effects of the polluted environment that are greenhouse effect, acid rain and serious health issues.

So, all the referred arguments make the biodiesel an only-way solution that will replace the diesel oil as long as it is able to respond to the current increased needs of the means of transport and has also the diesel oil characteristics so as to meet the requirements of producing the appropriate energy. Its main features are to be mixable, efficient and stable unexceptionally [6]. Micro emulsions, thermal cracking (pyrolysis), transesterification (alcoholysis) are the basic four proper procedures of producing biodiesel and according to the usual course of things the direct utilization and mixing follows [7].

At this point we should refer that the transesterification of natural and fats oils is the method that is mainly used. So the application of alkalis, acids or enzymes facilitates the process since they implement the catalysis to the transesterification chemical reaction of triglycerides with alcohols (ethanol, propanol, methanol, amyl alcohol or butanol).

The researchers [8] attempt to succeed in defining the accurate new fuels characteristics based on the extracted equations by using the linear regression model which is the basic tool for econometrics. According to their findings, the performance of socially responsible firms is negatively related to an increase of global CO2 emissions. In addition, the methods of econometrics will be applied to the variables targeting to retrace the connection points between the mix and the fuels physicochemical properties in the tested models.

In other words with the use of econometrics methods they to prove that there is indeed strong rates volatility and correlation between physicochemical properties of oils with their mixes. This relationship can be represented by mathematical equations after the application of linear regression [9, 10].

Naturally, as far as mixes are concerned, the Y = ax + b is the suggestive linear equation that is capable of pointing variables reactions. The variables relation could be stated by the simple linear regression model as the following equation shows:

$$Y_i = b_0 + b_i X + U_t \tag{1}$$

where Y_i: i is the experimental value of the dependent variable physicochemical property as well.

X_i: i is the value of the independent variable Mix.

b_0, b_1: the straight regression coefficients.

U_t: the equation error.

Finally, the referred estimating process should be applied to fuels identified with similar properties under the advisable temperature that it must be followed at its experiment. The use of econometrics in order to study the physicochemical properties of the oil enables further investigation of the relationship between physicochemical properties in the same or different oil blends. Through the equations it possible to identify in advance where to be expected the change of values of the mixtures and to preview the values of the examined physicochemical properties.

The process of experimenting followed international standards as well as ISO procedures in order for the results to be right. By using econometric methods, the authors have demonstrated that there is a positive relationship between the model variables. The extraction of equations was the main econometric result of the authors' research.

2. Materials

Biodiesel is one of the most well-known and widely used fuels in the world [23]. Speaking about oil and specific for Diesel oil we must mention that in in general is any liquid fuel used in diesel engines. Essentially diesel fuel is a mixture of hydrocarbons which are come from petroleum. Petroleum crude oils are composed of hydrocarbons of three major classes: (1) paraffinic, (2) naphthenic (or cycloparaffinic), and (3) aromatic hydrocarbons. The most common types of Diesel fuels are the biodiesel, biomass to liquid or gas to liquid Diesel.

When we are using the word Biodiesel we first have to give the definition of a particular fuel type [24]. Biodiesel refers to clean burning renewable fuel made using natural vegetable oils and fats. Biodiesel can be used as an alternative solution of the petroleum, diesel fuel or in most cases can be blended with petroleum diesel fuel in any proportion. Of course, this particular type of fuel is used not only on engines but is it also applicable to other uses.

It is known that the process by which the biodiesel is produced involves the esterification of the mother oil (and/or fat) with methanol and with the catalyst. The result may include components such as the residual catalyst, which are not desirable. For example, the presence of glycerol, although separated during the production of biodiesel, is almost certain in the final biodiesel.

Jet Propellant 8 (JP8) is a kerosene based fuel which when blended with specific additives constitutes a suitable material for military applications. This type of fuel has previously been used as a fuel for aircraft. However, its capabilities have also led to its use for ground vehicles [11].

The main reason was that military forces like NATO and especially the United States of America decided to use a specific fuel for their transport and activities. JP8 consists by approximately 99.8% kerosene by weight and is a complex mixture of higher order hydrocarbons, including alkanes, cycloalkanes, and aromatic molecules. JP8 contains three mandatory additives: a fuel system icing inhibitor, a corrosion inhibitor, and a static dissipater additive.

It is also known that JP8 is a fuel that can easily be adapted to any requirements and applications. This is done by adding chemicals with antioxidants.

Its composition is a mixture of petroleum hydrocarbons. In general, the mixture is of the methane series and contains 10–16 carbon atoms per molecule. Of course, the presence of paraffin is a key ingredient in making it used as a fuel in jet engines.

A framework for analyzing unstructured data using statistical methods in order to verify the existence and the type of correlation between the blends pure Diesel and Biodiesel and the respective physicochemical property.

It is also noted, in nowadays, that the statistics science is based on the chemistry and chemical engineering issues, so the interconnection between the statistical models and methods and the chemical experiments is very obvious provided that the chemical experiments and data analysis use them as a guide in the framework of their researches.

Firstly, a significant presentation of this interrelation should take place, and then the calibration matter should be studied. Speaking specifically, the generalized standard addition method (GSAM) will be examined targeting the following areas: initially, the upgrading of statistics field and the future estimation theory and the multicomponent analysis development as well and secondly the registration of the statisticians character in order to achieve a constitutive communication and understanding between the sciences and scientists of the two fields.

During our research we made experiments in order to export a regression equation which describes the variation of the physicochemical property taking into account, in most cases, the parameter of time.

Throughout our experiments we studied the variation of humidity of Diesel fuel with Time [12]; the variation of kinematic viscosity in blends of Diesel fuel with Biodiesel [4]; the variation of humidity and reduction conductivity of JP8 (F34) with time [13]; the variation of heat of combustion in blends of Diesel fuel with Biodiesel [14]; the variation of density in blends of Diesel fuel with Biodiesel [15]; the variation of conductivity in blends of Diesel fuel with Biodiesel [8]; the contribution of different kinds of biodiesel on the conductivity and density of blends of diesel and biodiesel fuels (animal and vegetable) [16]; the variation of humidity of conventional diesel with time [17]; the variation of density of diesel-biodiesel blends across the scale (0–100)% by adding each time 2% biodiesel and then measuring the density of 3 different temperatures (5, 15, 25)°C [9]. The results of our research are based on tables and figures. All tables and related figures are available in the published papers.

3. Results and discussion

In order to determine the suitability of fuel and the reduction of pollutants to the environment, quality control of alternative fuels is considered to be necessary. What we can certainly find almost all of the fuel is the presence of water. In any case, its presence causes problems such as erosion. Our effort is to modify the percentage of dampness in diesel oil through an equation. The whole process involves introducing hydrophilic LPG polymers into diesel fuel samples. Monitoring the whole process resulted in the moisture content being recorded over a period of time, and based on these results, we created an equation. In all cases and throughout our experiment the volume of fuel and mass remain constant [18–21].

Specifically, in this paper we have examined the regression model that we can see below,

$$Y_i = b_0 + b_i \, Sin \, (T_i) + U_t \tag{2}$$

The adaptability and suitability of the prototype model was confirmed by the application of specific econometric controls. Specifically, we examined the adaptability of the standard residues and their squares in order to check whether or not they are free from serial correlation. We also tested the existence of first order autocorrelation using the Durbin-Watson index [22].

The time (T) and the change in humidity was a basic question that we wanted to answer with this research using the linear regression model. The results showed that the presence of moisture declined over time, even reaching its maximum value in the first hour. Essentially with this equation we can predict humidity values over a period of time. By applying this equation we can know when we can remove the greatest amount of moisture from the fuel.

In our next research we tried to export an equation which describes the variation of kinematic viscosity in blends of diesel fuel with biodiesel. Using specific volume of these blends we determine kinematic viscosity via method ASTM D 445-06 using a capillary glass viscometer in order to study the contribution of quantity of biodiesel and convert the statistical data into mathematic relation as a specific formula, attempting to achieve an empirical evaluation.

Trying to accomplish this, we studied the way how the values of variables are changed and whether a relation exist using dispersion diagrams [4]. From the graphic depiction realized that the relation is linear and they proceeded to regression analysis. The analysis extracted the conclusion that the relation was strong and the values of the dependent variable kinematic viscosity was depended on a large percentage of the values of the mixture of fuels.

As far as the mix is concerned, the authors expected that the upgrade of the animal and vegetable biodiesel would be proportional to the increase of the kinematic viscosity arithmetic outcome due to the linear relation that connects the two variables. Studying on the regression, the authors notice that Eq. (4) has correct signals as they are aware of theory, so that the bigger addition of biodiesel on pure diesel will make the mix quote lower and lower real values to kinematic viscosity. The authors came to the relative conclusion as it is defined by the following: $Y_i = b_0 + b_i X + U_t$.

Y_i: i is the observation of the dependent variable kinematic viscosity;

X_i: i is the observation of the independent variable mix;

b_0, b_1 are the straight regression coefficients and U_t represents the equation error.

$$Y_i = 40.914 + 0.0540 X \qquad (3)$$

We come to the conclusion that there is an actual strong linear relation that connects the above mix with the kinematic viscosity. It is remarkable that the mix independent variable has a strong linear relation to the Kinematic Viscosity. The autocorrelation heteroskedasticity absence that was combined to the strong linear relation of the variables made the authors to conclude that Eq. (3) is able to preview the kinematic viscosity of the model fuels [4].

In 2012, Tsanaktsidis et al. [13] showed that the proposed equation can predict the moisture content and the height of conductivity in JP8 (F34) fuel over time. Essentially an ingredient, called hydrophilic polymer, leads to the reduction or even the elimination of humidity over a certain time. For example over 39% of humidity and 36% conductivity are decreased over 2 hours. The elimination of humidity makes the fuel suit able for car machines and gives combustion with less pollution for the environment. Thus, the quality of the fuel as well as its combustion efficiency can be improved while the reduction of water concentration enhances the secure of the combustion machine's operation.

The preliminary statistical analysis of sequences of humidity Y and time T have shown that the modifications of Y, probably determined by the sine of T. Specifically, in this paper we have examined the regression model in order to investigate the relation between humidity of JP8 fuel and the time, when the volume of the fuel and the mass of the polymer maintain stable. Some diagnostic tests were performed to establish goodness of fit and appropriateness of the model. First, the authors examined whether the standardized residuals and squared standardized residuals of the estimated model were free from serial correlation. In addition, the independence of the standardized residuals was confirmed by the Durbin Watson statistics.

$$Y_i = b_0 + b_i \, Sin\,(T_i) + U_t \tag{4}$$

More precisely, the results of the current investigation showed that the change of humidity could be described via a regression equation at which the dependent variable is the humidity change and the independent variable is the cosine of the time that the hydrophilic polymer TPA remains at the fuel. The results of the analysis showed that in the long run the presence of moisture decreases. In fact, it reaches its maximum value in the first hour. Hence, with the proposed equation, moisture rates can be predicted for a period of time. Eliminating work makes fuel more suitable for machines and polluting less the environment. Thus, the results of the present study give the possibility of humidity removal via the polymer TPA, which can be reduced [13].

Continuing our research and using specific volume of blends of Diesel fuel with Biodiesel we determined heat of combustion, in order to study the contribution of each kind of Biodiesel and we converted the statistical data into mathematic relations as a specific formula, attempting to achieve an empirical evaluation. The linear regression model was selected in order to correctly evaluate the values of the dependent variable (relative to the corresponding values of the independent variable). The preliminary statistical analysis of our series has led to the conclusion that the relation that links the variables of our model is linear [14].

We have created two blends which were blend 1, which referred to the pure diesel with the animal fats biodiesel and the blend 2 which referred to the pure diesel and the vegetable biodiesel. The two blends were studied in terms of the heat of combustion value in order to find the equation that might define the blends heat of combustion value. Studying on the regression, we notice that the Eqs. (5) and (6) had correct signals as the blend value increase should quote positively the heat of combustion variable.

$$Heat \; of \; Combustion \; 1 = \; 40.914 + 0.0540 \; blend \; 1 \tag{5}$$

$$Heat \; of \; Combustion \; 2 = \; 39.297 + 0.062 \; blend \; 2 \tag{6}$$

By studying the relationship of the variables initially with diagrams, we have concluded that the relationship between fuel (Diesel + Animal Fat Biodiesel = Mixture 1 and Diesel + Vegetable Biodiesel = Mixture 2) is linear. The strong linear positive relation of the variables made us conclude that the Eqs. (5) and (6) were able to preview the heat of combustion of the model fuels [14].

Furthermore Tsanaktsidis et al. [15] tried to export equations that would be able to describe the variation of density in blends of Diesel fuel with Biodiesel. Using specific volume of those blends, we determined density (while temperature maintain stable), in order to study the contribution of each kind of Biodiesel and we converted the statistical data into mathematic relations as a specific formula, attempting to achieve an empirical evaluation.

In that study [15] we used pure Diesel and two kinds of Biodiesel; Biodiesel by vegetables (vegetable oil fuel) and Biodiesel by animal fats. The total volume of each blend was 100 mL and in every measurement the volume of Biodiesel was changing. In the third blend the percentage of each kind of Biodiesel was 50% of the total volume of Biodiesel (for example 20% blend included: 80 mL Diesel, 10 mL Vegetable Biodiesel and 10 mL Animal Fats Biodiesel).

We measured density via the method ASTM D 1298-99 firstly in pure Diesel and then in all blends. The preliminary statistical analysis of the density series and the mix have showed that the variations of the variable den (Y) were defined by the rates of the variable mix (X) introducing linear equation for the three mix types that we took into account. We observe that Eqs. (2) and (4) has the appropriate signals as the mix rate increase should give plus rate to density variable.

The examination of the relation of the Diesel pure fuel to a specific mix ratio of fats and vegetable oils Biodiesel was implemented under stable climate conditions that were measured per mix in order to avoid the mistake risk of sample rate output. Eq. (7) that refers to the 1st mix (pure Diesel and vegetable oils biodiesel) attempts to explain the relation between the pure Diesel and the vegetable oils biodiesel, that were the Y and the X accounting the chemicals attributes of the variables that they express. We observed that the Eqs. (7) and (8) had the appropriate signals as the mix rate were increased should gave plus rate to density variable.

$$Density = 0.817461 + 0.00521 \ mix1 \ or \ Y = 0.817461 + 0.00521 \tag{7}$$

With reference to the 2nd mix (pure Diesel and vegetable oils biodiesel), the equation had the appropriate signals and the relation to the dependent variable were statistically significant for the 2nd fuel mix, on account of the reasons we have already referred.

$$Density = 0.830749 + 0.000312 \ mix2 \ or \ Y = 0.830749 + 0.000312 \tag{8}$$

For the 3rd mix (pure Diesel and aggregation of fats and vegetable oils biodiesel), the next equation were registered, which explained the linear relation of the model variable.

$$Density = 0.826332 + 0.000417 \ mix3 \ or \ Y = 0.826332 + 0.000417 \tag{9}$$

Based on the Eqs. (7), (8) and (9) we shown with can said that the dependent variable was statistically significant to each of the three mixes we have studied.

Taking into account the chemical features of the diesel fuels, fats and vegetable oils Biodiesel, and considering that the observation temperature is 15°C, we conclude that we might have high density predictability for all the mix types that we used [15] meaning the 1st the 2nd and the 3rd one. The equation application might only achieve quite perfect success in predicting the suitability of the final fuel in terms of the every time examined characteristic.

In our next research we tried to export a regression equation which describes the variation of conductivity in blends of Diesel fuel with Biodiesel. Using specific volume of those blends, we determined conductivity, in order to study the contribution of each kind of Biodiesel and we converted the statistical data into mathematic relations as a specific formula, attempting to achieve an empirical evaluation [8].

We used pure Diesel and one kind of Biodiesel; Biodiesel by Vegetables (soybean oil- vegetable oil fuel). Those samples met the specifications of Diesel fuel and Biodiesel Standards. The total volume of each blend was 100 mL (for example 20% blend included: 80 mL Diesel and 20 mL Vegetable Biodiesel) and in every measurement the volume of Biodiesel was changing.

Studying on the regression, we noticed that the Eq. (10) had correct signals. The bigger addition of Biodiesel on pure Diesel would make the mix quote lower and lower real prices to the Conductivity.

$$Conductivity = 1691.409 - 17.504 \ mix \ or \ Y = 1691.409 - 17.504 \tag{10}$$

In the framework of our research on the fuels relation (diesel + biodiesel = blend or mix), we concluded that there were an actual strong linear relation that connects the above mix with the Conductivity. It was considerable that the mix independent variable has a linear relation to the Conductivity.

That model meaning Eq. (10) probably might give not give us secure conclusions because of the substantial part of the humidity in the conductivity factor. The humidity factor importantly influences the conductivity values so that an additional measurement will ensure the inferences in this research field.

One of the filed that needs more attention and further study is the contribution of different kinds of biodiesel on the conductivity and density of blends of diesel and biodiesel fuels (animal and vegetable). In the framework of this research we attempted to substantiate the existence and the type of the correlation between the blends of pure Diesel and Biodiesel (Vegetable and Animal Biodiesel) and the Density as well as the conductivity.

We decided to adopt for this purpose the Regression Analysis as the best model that would support this study. The basic target of this method was the potentiality of the precise evaluation of the dependent variable prices regarding the specific independent variables prices. The preliminary statistical analysis of the Density (Y_1) and the Conductivity (Y_2) with respect to the fuel volume fractions (X) (fuels) series has directly showed that the relation that connects each variable is linear [16].

Pure diesel and biodiesel (vegetable and animal) were used. Those samples met the specifications of diesel fuel and biodiesel standards described above. Eleven different blends were used. The total volume of each blend was 100 mL and in each blend the volume fraction of diesel/biodiesel was different (100% diesel, 90% diesel-10% biodiesel, 80% diesel-20% biodiesel, 70% diesel-30% biodiesel, 60% diesel-40% biodiesel, 50% diesel-50% biodiesel, 40% diesel-60% biodiesel, 30% diesel-70% biodiesel, 200% diesel-80% biodiesel, 10% diesel-90% biodiesel, and 100% biodiesel).

In this research we had six different blends of biodiesel. The mixes were studied in terms of their Conductivity and Density values in order to find an equation that shows the relationship between Conductivity and Density of the created mix as a function of the blend's constitution. As far as the produced mix is concerned, the "mix", our last assumption is that the upgrade of biodiesel (vegetable or animal or 50% vegetable with 50% animal biodiesel) would be proportional to the increase in Density and Conductivity due to the numerical result of a linear relationship between the two variables.

$$Density = 0.005mix + 0.8175 \quad Y_1 = 0.005mix + 0.8175 \tag{11}$$

$$Conductivity = 844.55\,mix \;\; - 83.63 \;\; or \;\; Y_2 = 844.55mix - 83.63 \tag{12}$$

The tested model was statistically significant at 1% level and we came to the decision that the selected regression model for the above mixes were suitable to account for an important part of the Density and Conductivity variability. We proceed to a closer examination of blends of pure diesel with animal biodiesel. The mixes were studied in terms of the Conductivity and the Density. As far as the produced mix were concerned, the "mix", we expected that the upgrade of animal biodiesel would be proportional to the increase in Conductivity and also in Density due to the numerical result of a linear relationship between the two variables.

$$Desity = \;\; 0.004mix + 0.8215 \;\; or Y_1 = 0.004mix + 0.8215 \tag{13}$$

$$Conductivity = \;\; 2.789mix - 29.364 \;\; or \;\; Y_2 = 2.789mix - 29.364 \tag{14}$$

The tested model was statistically significant at 1% level and we came to the decision that the selected regression model for the above mixes were suitable to account for an important part of the Density and Conductivity variability.

Throughout the rest of that paper [16], we continued to our analysis with the examination of blends of pure diesel with Biodiesel (vegetable 50% + animal 50% biodiesel). Due to the linear relationship between the two variables our regression analysis has shown that the Eqs. (15) and (16) had correct signals.

$$Density = 0.005mix + 0.8149 \;\; or \;\; Y_1 = 0.005mix + 0.8149.$$
$$Mix = Pure \;\; Diesel + Biodiesel \;\; (50\% Animal \;\; Biodiesel + 50\% Vegetable \;\; Biodiesel \tag{15}$$

$$Conductivity = 11.605mix - 178.4 \;\; or \;\; Y_2 = 11.605mix - 178.64 \;.$$
$$Mix = Pure \;\; Diesel + Biodiesel \;\; (50\% Animal \;\; Biodiesel + 50\% Vegetable \;\; Biodiesel) \tag{16}$$

The tested model was statistically significant at 1% and based on that procedure we are able to come to secure conclusions. Moreover with this procedure it is possible the further study of the biodiesel use with lower density and Conductivity. We consider that this scientific study can contribute to the today's industry sector in terms of the exploitation of the alternative biodiesel fuel. Finally due to fact that the biodiesel cost is lower than the pure diesel the utilization possibility in a wide range reduces the production cost and makes a final fuel product that is friendlier to the environment [16].

In our next research we studied the possibility of exporting a regression equation which could describe the variation of humidity of conventional diesel with time. A hydrophilic polymers TPA (Thermal Polyaspartic Anion) and natural resin from halepius Pines tree were used to eliminate humidity from conventional diesel. At both cases, where TPA as well as natural resin was used as additive, the hydrophilic polymers just blended, mechanically, with the diesel and after several mixing times were removed from this. The elimination of humidity made the fuel suitable for car machines and gave combustion with less pollution for the environment [17].

The preliminary statistical analysis of sequences of humidity Y and time T have shown that the modifications of Y, probably determined by the sine of T. Specifically, in this research we have examined the regression model

$$Y_i = b_0 + b_i \, Sin \, (T_i) + U_t \tag{17}$$

in order to investigate the relation between humidity of JP8 fuel and the time, when the volume of the fuel and the mass of the polymer maintain stable.

As far as for the statistical investigation for the variables like time T and resin we came to the conclusion that the relationship is not linear. So, in order to proceed our research we have examined the regression model $Y_i = b_0 + b_1(T_i) + b_2(T_i)^2 + u_t$ in order to investigate the relation between humidity (Y), Time (Ti) and Time 2 (Ti) and the result of our analysis have shown the following equation:

$$Humidity \left(\frac{mg}{g} \right) = 137.58 + 1.33 Time + 0.004 Time^2 \tag{18}$$

The results of the current investigation have shown that the change of humidity could be described via a regression equation at which the dependent variable was the humidity change and the independent variable was the cosine of the time that the hydrophilic polymer TPA remains at the fuel.

Thus, the results of this research [17] gave the possibility of humidity removal via the polymers TPA and RESIN. Hence the quality of the fuel as well as its combustion efficiency can be improved while significant problems can be avoided because of the presence of water in the combustion machine.

Moreover, the properties of the fuel were not influenced by the use of the polymer. Via the equation, the value of humidity in a fuel can be calculated in the frames of the experiment time scale, without the use of experimental process, only by maintaining the parameters of the experiment (temperature 25°C and polymer use) in the proposed proportion.

The present study also investigated how fuel's humidity changes by the time (T) that the resin mass remains at the fuel, using a regression model. Because of the nonlinear relation we can say with certainty that the humidity can be predicted with safety (almost 86% at the time) when using Eq. (18).

The use of biodiesel fuel is becoming increasingly imperative nowadays and it is necessary to know the change of density. In our next research we have studied the variation of density of diesel-biodiesel blends across the scale (0–100)% by adding each time 2% biodiesel and then measuring the density of three different temperatures (5, 15, 25)°C covering and the usual scale of temperatures the use of mixtures of diesel-biodiesel. Through the extraction of equations can be known in advance the relationship of density of diesel-biodiesel blend, and temperature that is used.

Based on these fuels created 50 diesel-biodiesel blends content (0-2-4-6 .... 100) % v/v, at three different temperatures (5, 15, 25)°C to cover all common temperature scale used diesel-biodiesel mixtures. Then we proceed to the determination of the density of these mixtures. The determination is conducted through the ASTM D-1298 method (ASTM D1298-99, 2005) with measurements by means of BS718:1960LSOSP hydrometers. These measurements are reduced to a temperature of 15°C, at which they also constitute the value of fuel density, while they are expressed in kg/L. The measurement scale of these hydrometers is between 0.6 and 1.1.

In order to verify the existence and the type of correlation between the blends pure Diesel and Biodiesel and the Density we decided to use the Regression Analysis as the best method that would support this study. The preliminary statistical analysis of the Density (Y) with respect to the fuel volume fractions (X) (fuels) series had directly showed that the relation that connects each variable were linear for its temperature meaning 5, 15 and 25°C [9].

In this research we had almost 50 different blends of biodiesel for its temperature. The mixes were studied in terms of their Density values in order to find an equation that shown the relationship between Densities of the created mix as a function of the blend's constitution. As far as the produced mix was concerned, the "mix", our last assumption was that the upgrade of biodiesel would be proportional to the increase in Density due to the numerical result of a linear relationship between the two variables. We came to the relative conclusion as it was defined by the following equations:

$$mix01 = 0.8114 + 0.0013 \ density \ \text{Constant temperature 5°C} \qquad (19)$$

$$mix02 = 0.8282 + 0.001 \ density \ \text{Constant temperature 15°C} \qquad (20)$$

$$mix03 = 0.8262 + 0.0008 \ density \ \text{Constant temperature 25°C} \qquad (21)$$

The proposed methodology can be used in the bio fuels industry for the prediction of variation of the density of mixtures of diesel-biodiesel in a temperature scale is the most common in use for these fuels. Moreover based on this procedure we are able to come to secure conclusions when the values of density according to the mixes where measured between 5 and 25°C. If we overcome these limits then we will face autocorrelation and heteroskedasticity problems. In this case our model will have no predictive ability and will essentially reject as unacceptable [9].

4. Conclusion

Based on our results the variation of the physicochemical properties of the oil can be predicted. This can be done using the equations generated during our investigations. The predictive capacity of these equations is valid only if specimens and mixtures follow specific rules, such as those during the experiments we conducted. With these studies we came to the conclusion that it is given the opportunity to develop software in order to study the changes of physico-chemical properties of petroleum distillate products. The development of such an application would help us to know in advance the variation of the physicochemical properties. This implementation would be important not only for researchers but also for the respective control bodies as regards the quality of the final product at each stage to the final consumer. Taking into account such equations and having knowledge of oil technology, we can predict the prices per fuel mix and, accordingly, accept it or reject it.

Acknowledgements

At this point we would like to stress that without the use of the facilities of Technological Education Institute of Western Macedonia and specific the laboratory of Qualitative Fuel Control ISO 9001: 2008 this study would not be possible.

Author details

Constantinos Tsanaktsidis[1]* and Konstantinos Spinthiropoulos[2]

*Address all correspondence to: prof.tsanaktsidis@gmail.com

1 Technological Education Institute of Western Macedonia, Department of Pollution Control and Technologies, Kozani, Greece

2 Technological Education Institute of Western Macedonia, Department of Accounting and Finance, Kozani, Greece

References

[1] Samuelson PA, Koopmans TC, Stone JRN. Report of the evaluative committee for econometrica. Econometrica. 1954;**22**:141-146

[2] Samuelson PA, Nordhaus WD. Economics. 18th ed. Vol. 5. NY: McGraw-Hill; 2004

[3] Available from: http://www.dziejekrakowa.pl/biogramy/index.php?id=516

[4] Tsanaktsidis CG, Spinthiropoulos KG, Christidis SG, Basileiadis VM, Garefalakis AE. Production of a mathematic equation using statistical data for the determination of kine-matic viscosity in blends of diesel fuel with biodiesel. Computer Technology and Applica-tion. 2012;**3**:393-399

[5] Greene D. Motor fuel choice: An econometric analysis?. TWLC~II. Ru.-A. vd. DA. 1989;**3**: 3-2.53. 1989 Rimed in Great Britain

[6]   Ramesh D, Samapathrajan A, Venkatachalam P. Production of Biodiesel from *Jatropha curcas* Oil by Using Pilot Biodiesel Plant. India: Agrl. Engg. College & Research Institute; 2002

[7]   Ma F, Hanna AM. Biodiesel production: A review. Bioresource Technology. 1999;**70**:1-15

[8]   Tsanaktsidis CG, Christidis SG, Spinthiropoulos KG, Tzilantonis GT. Exporting a regression equation for the determination of conductivity in blends of diesel fuel with biodiesel. In: Proceedings from CISSE'12: International Joint Conferences on Computer, Information, Systems Sciences and Engineering. USA: University of Bridgeport; 2012

[9]   Tsanaktsidis C, Spinthiropoulos K, Tzilantonis G, Katsaros X. Variation of density of diesel and biodiesel mixtures in three different temperature ranges. Petroleum Science and Technology. 2016;**34**(13):1121-1128

[10] Tsanaktsidis CG, Christidis SG, Tzilantonis GT. Study about effect of processed biodiesel in physicochemical properties of mixtures with diesel fuel in order to increase their antifouling action. International Journal of Environmental Science and Development. 2010;**1**:205-207

[11] Rakopoulos CD, Hountalas DT, Rakopoulos DC, Levendis YA. Energy & Fuels. 2004;**18**: 1302-1309

[12] Tsanaktsidis CG, Sariannidis N, Christidis SG. Regression analysis about humidity elimination from diesel fuel via bioorganic compounds to increase antifouling action. In: Proceedings of International Joint Conferences on Computer, Information, and Systems Sciences, and Engineering (CISSE 09); 4-12 December 2009; Vol 1: Technological Developments in Networking, Education and Automation. USA; 2010. pp. 377-385

[13] Tsanaktsidis CG, Sariannidis N, Christidis SG, Itziou A. Regression analysis about humidity elimination and reduction conductivity from JP8 via a hydrophilic polymer. Petroleum Chemistry. 2012;**52**:447-451

[14] Tsanaktsidis CG, Vasileiadis VM, Spinthiropoulos KG, Christidis SG, Garefalakis AE. Statistical analysis to export an equation in order to determine heat of combustion in blends of diesel fuel with biodiesel. Lecture Notes in Electrical Engineering. 2013;**152**:719-726

[15] Tsanaktsidis CG, Spinthiropoulos KG, Christidis SG, Sariannidis N. Statistical analysis about variation of density in blends of diesel fuel with biodiesel. Chemistry and Technology of Fuels and Oils. 2013;**49**:399-348

[16] Tsanaktsidis CG, Kiratzis N, Tzilantonis GT, Sariannidis N, Spinthiropoulos KG. Variation of density and conductivity with mixtures of diesel and biodiesel (animal and vegetable) by analysis of variance using the linear regression and interpretation using mathematical equations. In: 5th Annual International Conference on Sustainable Energy and Environmental Sciences (SEES 2016); 2016. pp. 111-118

[17] Tsanaktsidis CG, Sariannidis N, Spinthiropoulos KG, Christidis SG, Tzilantonis GT. Statistical analysis and comparative about humidity elimination in conventional diesel fuel

using synthetic and natural hydrophilic polymers as additives. Petroleum Science and Technology. 2016; in press

[18] Chuvieco E, Riaño D, Aguado I, Cocero D. Estimation of fuel moisture content from multitemporal analysis of Landsat Thematic Mapper reflectance data: Applications in fire danger assessment. International Journal of Remote Sensing. 2002;**23**:2145-2162

[19] Xu M, Fan Y, Yuan J, Sheng C, Yao H. A simplified fuel-nox model based on regression analysis. International Journal of Energy Research. 1999;**23**:157-168

[20] Carter D, Rogers DA, Simkins BJ. Does fuel hedging make economic sense? The case of the US airline industry. AFA 2004 San Diego Meetings; 2002

[21] Cebrat G, Karagiannidis A, Papadopoulos A. Proposing intelligent alternative propulsion concepts contributing to higher CO_2 savings with first generation biofuels. Management of Environmental Quality: An International Journal. 2008;**19**:740-749

[22] Durbin J, Watson GS. Testing for serial correlation in least square regression. Biometrika. 1950;**37**:409-428

[23] Azocar L, Ciudad G, Heipieper J, Navia R. Biotechnological processes for biodiesel production using alternative oils. Applied Microbiology and Biotechnology. 2010;**88**:621-636

[24] Daroch M, Geng S, Wang G. Recent advances in liquid biofuel production from algal feedstocks. Applied Energy. 2013;**102**:1371-1381

Biological Treatment of Petrochemical Wastewater

Nirmal Ghimire and Shuai Wang

Additional information is available at the end of the chapter

http://dx.doi.org/10.5772/intechopen.79655

Abstract

Petrochemical wastewater is inherent to oil industries. The wastewater contains various organic and inorganic components that need to be well managed before they can be discharged to any receiving waters. The complexity of the wastewater and stringent discharge limit push the development of wastewater treatment by combinations of different methods. Biological wastewater treatments that have been well developed for organic and inorganic wastewater treatment are thus a potential method for petrochemical wastewater management. This chapter summarizes the commonly applied petrochemical wastewater pretreatment methods prior biological treatments and compares different biological treatment systems' performance such as biological anaerobic, aerobic and integrated systems. Two case studies are presented for a high chemical oxygen demand (COD) contents petrochemical wastewater treatment in full-scale by applying Biowater Technology's biofilm system continuous flow intermittent cleaning (CFIC) and a pilot-scale study by an integrated anaerobic and aerobic biofilm system hybrid vertical anaerobic biofilm (HyVAB). Both processes showed substantial (over 90%) COD removal, while the HyVAB system produced high methane content biogas that can be potentially used as an energy source. Studies of degradation of certain toxic chemicals, such as aromatic compounds in petrochemical wastewater, by the advanced treatment systems incorporating specific organisms can be of good research interest.

Keywords: petrochemical wastewater, anaerobic digestion, aerobic digestion, biofilm reactor, integrated system

1. Introduction

Increasing consumption of oil in modern society has led to more oil/oil refinery waste generation. The oil processing wastewater/waste has high concentrations of aliphatic, aromatic petroleum

IntechOpen

hydrocarbons, etc. Direct discharge of this will affect plants and aquatic life of surface and ground water sources. Due to its organic origination, complex nature, and toxic effects, wastewater treatment prior to discharge is obligatory. The biological treatment process is normally applied to reduce the effects of petrochemical waste.

Stringent regulations have motivated researchers to design advanced treatment facilities to give high treatment efficiency, low maintenance, footprint, and operational costs. Biological anaerobic, anoxic, and aerobic digestion (or a combination of each other) have been implemented to treat petrochemical wastewater. Optimizing pretreatment process using physicochemical processes is also important for getting suitable pretreatment wastewater for efficient biological secondary treatment. An overview and update of the petrochemical wastewater treatment processes will contribute to the knowledge development both theoretically and practically.

In this section, the petrochemical wastewater treatment by biological processes is shortly reviewed and discussed. Section 2 introduces the petrochemical wastewater sources and their components in general. Section 3 introduces the normally applied pretreatment process prior to biological treatment processes. Section 4 presents the commonly applied anaerobic, aerobic, and combined anaerobic and aerobic biological systems for petrochemical wastewater treatment. Section 5 shows two case studies on the petrochemical wastewater treatment using Biowater Technology AS's continuous flow intermittent cleaning (CFIC) and hybrid vertical anaerobic biofilm (HyVAB) processes. Section 6 summarizes challenges and further studies in the petrochemical wastewater treatment.

2. Petrochemical wastewater

Petrochemical wastewater is a general term of wastewater associated with oil-related industries. The sources of petrochemical wastewater are diverse and can originate from oilfield production, crude oil refinery plants, the olefin process plants, refrigeration, energy unities, and other sporadic wastewaters [1, 2]. The compositions of wastewater from different sources consist of varying chemicals and show different toxicity and degradability in terms of biological treatment. In this chapter, to better compare the treatment efficiency with varying pretreatment processes, the petrochemical wastewater has been categorized to oilfield-produced wastewater, petrochemical refinery, and oily wastewater based on the originates.

Oilfield-produced wastewater is generated in crude oil extraction from oil wells that contain high concentrations of artificial surfactants and emulsified crude oil characterized of high COD and low biodegradability [3]. It is produced during oil extraction in oil fields and contains complex recalcitrant organic pollutants such as polymer, surfactants, radioactive substances, benzenes, phenols, humus, polycyclic aromatic hydrocarbons (PAHs), and different kinds of heavy mineral oil [4, 5]. **Table 1** presents the commonly found compositions of wastewater obtained from oilfield production.

Petroleum refinery wastewater is generated in oil refinery processes that produce more than 2500 refined products. The wastewater can be from cooling systems, distillation, hydrotreating, and

Parameter	Values	Heavy metal	Values (mg/L)
Density (kg/m³)	1014–1140	Calcium	13–25,800
Surface Tension (dynes/cm)	43–78	Sodium	132–97,000
TOC (mg/L)	0–1500	Potassium	24–4300
COD (mg/L)	1220	Magnesium	8–6000
TSS (mg/L)	1.2–1000	Iron	<0.1–100
pH	4.3–10	Aluminum	310–410
Total oil (IR; mg/L)	2–565	Boron	5–95
Volatile (BTX; mg/L)	0.39–35	Barium	1.3–650
Base/neutrals (mg/L)	<140	Cadmium	<0.005–0.2
(Total non-volatile oil and grease by GLC/MS) base (g/L)	275	Chromium	0.02–1.1
Chloride (mg/L)	80–200,000	Copper	<0.002–1.5
Bicarbonate (mg/L)	77–3990	Lithium	3–50
Sulfate (mg/L)	<2–1650	Manganese	<0.004–175
Ammoniacal nitrogen (mg/L)	10–300	Lead	0.002–8.8
Sulfite (mg/L)	10	Strontium	0.02–1000
Total polar (mg/L)	9.7–600	Titanium	<0.01–0.7
Higher acids (mg/L)	<1–63	Zinc	0.01–35
Phenols (mg/L)	0.009–23	Arsenic	<0.005–0.3
VFA's (volatile fatty acids) (mg/L)	2–4900	Mercury	<0.001–0.002
		Silver	<0.001–0.15
		Beryllium	<0.001–0.004

Table 1. Wastewater parameter form oilfield production [6].

desalting. The compositions of the refinery wastewater can vary depending upon the operational units for different products at specific time and locations. Different concentrations of ammonia, sulfide, phenols, Benzo, and other hydrocarbons are normally present in such wastewater [7, 8].

The oily wastewater is defined here to be any wastewater that does not clearly belong to the two categories mentioned earlier. This wastewater can be from petrochemical-related indus-tries such as from oil transportation tank, garage oil wastewater, etc. The composition of such wastewater is diverse with high COD that can be over 15 g/L [9].

3. Pretreatment process for biological stabilization

Wastewater from petrochemical industries consists of different chemicals. The treatment pro-cesses depend and are specialized by wastewater sources, discharge requirements, and treatment efficiencies. Normally, pretreatment processes are applied in the treatment of petroleum refinery wastewater before it is sent to biological process for organic elimination [8]. A primary treatment includes the elimination of free oil and gross solids; elimination of dispersed oil and solids by flocculation, flotation, sedimentation, filtration, microelectrolysis, etc.; increasing the biodegradability of wastewater, etc. [8]. This chapter lists a few commonly applied methods for petrochemical wastewater pretreatment.

3.1. Physical treatment

Depending on the wastewater characteristics, physical treatment such as adsorption by active carbon, copolymers, zeolite, etc. can be used for removing hydrocarbons in the petrochemical wastewater [6]. Evaporation is proposed to remove oil residuals in saline wastewater. Dissolved air flotation (DAF) is commonly used for wastewater containing oil/fat as well as suspended solids, which can also be applied for petrochemical wastewater.

Microfiltration (MF) and ultrafiltration (UF) are also applicable for pretreatment before the wastewater passes through, for example, reverse osmosis (RO) process for reusing purposes [10].

3.2. Chemical treatment

Enhancing hydrolysis by adding chemicals for removing the long-chain organics, toxic material, or suspended solids can increase the Biochemical Oxygen Demand (BOD) ratio of the wastewater. Three chemical treatment processes are listed here.

Micro-aeration breaks down high hydrocarbon content components from wastewater, which leads to easily biodegradable organic generation. At a dissolved oxygen (DO) concentration from 0.2 to 0.3 mg/L, the hydrolysis of wastewater organics is enhanced. The BOD/COD ratio is increased and SO_4- reduction in wastewater is inhibited. Low H_2S generation due to SO_4- reduced reduction can benefit subsequent biological treatment by lowering inhibitory effects. Benzene ring organics', such as benzene, toluene, ethylbenzene, and xylenes, treatability in the biological stage can be improved [11].

Coagulation-flocculation for specific petrochemical wastewater treatment, such as purified terephthalic acid (PTA) production wastewater; the wastewater contains aromatic compounds such as p-toluic acid, benzoic acid, 4-carboxybenzaldehyde, phthalic acid (PA), and terephthalic acid (TA), etc. Ferric chloride is found to be the most effective coagulant with COD removal efficiency at 75.5% at wastewater COD of 2776 mg/L and dose of pH 5.6. Adding cationic polyacrylamide improves the sludge filtration [12]. Certain streams that combine coagulation and flocculation as pretreatment followed by MF and UF achieved significant suspended solid removal [10].

Ozonation for wastewater that contains phenol, benzoic acid, aminobenzoic acid, and petrochemical industry wastewater containing acrylonitrile butadiene styrene (ABS) at 30 min and 100–200 mg O_3/h showed an increased BOD/COD ratio from 20 to 35% [13].

3.3. Other treatment

Microelectrolysis of petrochemical wastewater has been tested with positive effects on the COD removal as well as increasing the BOD-to-COD ratio levels [14].

4. Biological treatment of petrochemical wastewater

Biological treatment incorporates actions of different microbes to eliminate organics and stabilize hazardous pollutants in petrochemical wastewater. Stringent environmental standards and recycling of water for reuse have shifted focus to biological treatments because of its cost and pollutant removal efficiency. As the nature of petrochemical wastewater is very complex, biological treatment to remove pollutants still has challenges despite immense potentials. Complex structures of aromatic, polycyclic, and heterocyclic ringed chemicals are known to be restraint to biological degradation [15]. However, recent research activities have produced notable removal percentages of pollutants from petrochemical wastewater [16].

Anaerobic digestion (AD), aerobic digestion, or an integration of both methods is commonly applied in biological processes to treat petrochemical wastewater.

4.1. Anaerobic process

Anaerobic digestion has the advantages of producing methane as a renewable energy, requiring less space and having lower sludge generation than aerobic process. A literature review of anaerobic digestion on the petrochemical wastewater is given in **Table 2**. Petrochemical wastewater treated in anaerobic baffled reactor (ABR), sequence batch, and up-flow sludge blanket reactor (UASB) was commonly applied. It shows that organics in the petrochemical wastewater could be partially anaerobic digested at a removal efficiency depending on the chemical constituents, reactor type, operational conditions (temperature, loading rate, etc.), and wastewater sources [24].

COD removal efficiency is used here as a general parameter to assess the performance of different systems. Crude oil extraction of light, medium, and heavy petroleum wastewater treatment by different anaerobic digestion systems at mesophilic or thermophilic conditions showed that in batch test over 56–71% COD removal was achievable at thermophilic condition [1, 18] (**Table 2**), while UASB system can achieve over 93% COD removal at mesophilic conditions for wastewater from light petroleum extraction (**Table 2**). It seems light petroleum extraction wastewater was generally easily degradable (over 71–93% removal) compared to the medium and heavy oil extraction wastewater. The setup of plug flow pattern and granular sludge application in UASB might also enhance the interaction between wastewater and organisms, giving higher efficiency. The removal efficiency decreases as the loading rate increases, indicating the inhibition effects to the organisms.

Medium- and heavy oil-produced wastewater treatment efficiency was relatively low. Batch system gives generally a better treatment efficiency for these two wastewaters at about 50–60%

NO.	Types of wastewater	Treatment system	Operating conditions	Pollutants monitored	Removal efficiencies (%)	References
1	Crude oil extractions*	Batch reactors	Thermophilic conditions (55 ± 1°C)	COD	70.7 59.9 62.1	[1]
		UASB	Mesophilic	COD	81.7 23.5 35.7	[17]
2	Crude oil extractions**	Batch reactors	Thermophilic conditions (55 ± 1°C)	COD	68.2–69.2 55.9–50.4	[18]
3	Crude oil extractions***	UASB	Mesophilic 1.06 kg COD/m³.d 0.78 kg COD/m³.d	COD	93 26	[19]
4	Crude oil extraction of light petroleum	UASB	Mesophilic 4.7 kg COD/m³.d 0.78 kg COD/m³.d	COD	23.8 86.1	[20]
		UASB	Mesophilic 5.6 kg COD/m³.d Thermophilic 5.6 kg COD/m³.d	COD VSS COD VSS	40–80 42–73 67–84 52–67	[21]
		UASB	Thermophilic 1.1 kg COD/m³.d	COD	78	[22]
		UASB	4.1 kg COD/m³.d	COD	82	[23]
5	Heavy oil refinery	UASB	3.4 kg COD/m³.d	COD Total oil	70 72	[9]
		ABR	0.5 kg COD/m³.d	COD Oil	65 88	[24]

*Water from light petroleum, medium petroleum and heavy petroleum, respectively.
**Water from medium petroleum and heavy petroleum, respectively.
***Water from light petroleum, medium petroleum, respectively.

Table 2. Overview of anaerobic treatment of petrochemical wastewater.

removal (**Table 2**), while UASB shows low efficiency at around 20–30% removal efficiency. The effects of toxic chemicals in the wastewater and high content of large organic molecules can be the reason for low efficiency.

4.2. Aerobic process

Aerobic process has been applied widely in petrochemical wastewater treatment attributed to its features of easy operation, less sensitiveness to toxic effects, higher organisms' growth rate, etc. than the anaerobic system. Different aerobic reactors such as traditional active sludge, contact stabilization active sludge, sequence batch reactor (SBR) that applies active sludge and biological aerated filter (BAF), membrane bioreactor (MB), moving bed biofilm reactor (MBBR), aerobic submerged fixed-bed reactor (ASFBR) that applies biofilm, etc. have been

S. N	Types of wastewater	Treatment system	Operating conditions	Pollutants monitored	Removal efficiencies (%)	References
1	Petroleum refinery	Contact stabilization	F/M 0.38	COD BOD NH$_3$-N H$_2$S TSS	97.9 95.8 87.5 97.5 98.6	[25]
		Activated sludge		COD BOD NH$_3$-N H$_2$S TSS	93.4 94.4 83.3 95 97.6	
		Activated sludge		COD TOC TSS	94–95 85–87 98–99	[8]
		SBR		COD TOC	80 84	[26]
		MSBR	SRT: 20 days HRT: 8 h	COD Oil and grease TPH	80 82 93.4	[27]
		HF-UF MBR	HRT: 25–36 h	COD TSS Turbidity	82 98 98	[28]
		CF-MBR	DO: 4 mg/L F/M: 0.2–1.15	COD	93–94	[29]
		BAF	1.9 kg COD/m^3.d	COD Oil SS	84.5 94 83.4	[30]
		ASFBR	2.4 kg COD/m^3.d HRT: 12 h	COD TSS	70±7 65±16	[31]
2	Oilfield	BAF with immobilized carriers	1.1 kg COD/ m^3.d	TOC Oil	78 94	[5]
		MBBR with Activated sludge	4.2 kg COD/ m^3.d	COD	74	[32]
		Activated sludge	SRT: 20 days MLSS: 730 mg/L	THP	98–99	[33]
		Airlift reactor	HRT: 12 days	COD TOC Phenols NH4 + -N	65 80 65 40	[34]
3	Oily wastewater	Activated sludge	Temperature: 25–37°C	COD Ethylene dichloride Vinyl chloride Total hydrocarbons	89 99 92 80	[35]
		Activated sludge and contact oxidation	1.1 kg COD/ m^3.d	COD NH4$^+$-N	84.9 60	[36]

S. N	Types of wastewater	Treatment system	Operating conditions	Pollutants monitored	Removal efficiencies (%)	References
		UF Membrane bioreactor	Temperature–35°C	COD TOC Oil	97 98 99.9	[37]
		RBC	Diesel concentration: 0.6%	TPH COD	98.1 97.2	[38]
		CFIC	Temperature–35°C	COD	92	Case study in chapter 5

Table 3. Overview of aerobic treatment process of petrochemical wastewater.

tested to treat petrochemical wastewater from varying sources and presented in **Table 3**. Generally higher COD and chemical removal efficiencies by aerobic process are achieved than the anaerobic processes (**Tables 2** and **3**). The sludge retention time, hydraulic retention time, dissolved oxygen level, feed to organism ratio, and temperature are some of the important factors that determine the treatment efficiency.

Petroleum refinery wastewater COD removal was generally high from 70 to 98% in the mentioned aerobic system (**Table 3**), which in anaerobic system is from 70 to 93%. The contact and extended active sludge process can achieve high COD removal rate of 89–95% (**Table 3**) at a feed to microorganism ratio of 0.38 [25]. The applied aeration to the mixed liquor and the sludge recycle rate was found to be critical parameters in the successful optimization of the contact stabilization process. The treatment efficiency of NH_4-N, H_2S, and TSS were also high [25]. Traditional SBR has relatively lower treatment efficiency at 80% COD removal (**Table 3**).

The membrane reactors such as BAF, cross-flow membrane bioreactor (CF-MBR), membrane sequencing batch reactor (MSBR), and hollow fiber ultrafiltration membrane bioreactor (HF-UF MBR) including ultrafiltration MBR systems treating higher OLR or food to organisms' ratio can achieve over 80% COD removal (**Table 3**). MBBR system applying biofilm can achieve 74% COD removal at a high OLR of 4.2 kg $COD/m^3 \cdot d$ (**Table 3**). It also can be seen that NH_4-N and H_2S removal are above 60% that cannot be obtained in anaerobic system. The Total Organic Compounds (TOC) and oil removal are also better than the anaerobic system.

Oilfield wastewater is relatively reluctant to aerobic digestion due to the complex ingredient. The removal efficiency of such water has a COD removal at around 30–74% (**Table 3**) by BAF, MBBR, etc. Active sludge process seems to handle well the wastewater and achieve high total petroleum hydrocarbon (TPH) removal.

The oily wastewater COD removal is generally high by using different aerobic methods, indicating its easily degradable nature (**Table 3**). The case study in Section 5 presents the advanced biofilm technology named CFIC process by Biowater Technology AS. The full-scale plant data show consistently high COD removal efficiency over 90%.

4.3. Integrated biological process

The treatment efficiencies of individual anaerobic and aerobic systems show good capability in treating certain petrochemical wastewater. An integrated system combining anaerobic and aerobic processes can possibly take the advantages of both and achieve even better removal efficiency for chemicals that are not easily degraded by either anaerobic or aerobic process. An integrated system that is focused in this chapter can be a hybrid reactor consisting of an anaerobic and an aerobic system in a vertical design, such as a hybrid vertical flow anaerobic aerobic biofilm reactor (HyVAB) [9], provided by Biowater Technology AS, or a combination of different treatment processes in series, for example, a system consists of traditional anaerobic reactor and an aerobic stage in series. The performance of integrated systems for petrochemical refinery, oilfield-produced wastewater, and other oily wastewaters is presented in **Table 4**. The integrated system could effectively remove easily degradable COD in the anaerobic stage first and convert it to biogas with the residual COD and other chemicals such as ammonium, sulfide, etc. degraded in the aerobic stage (**Table 4**).

Hybrid system combining UASB and aerobic stage treating oilfield wastewater showed good effects on COD removal by enabling acidification prior to the aerobic stage where organisms are actively reacting with organic chemicals. The COD removal rates were over 70–95%. Oil and ammonia removal was also recorded over 87% (**Table 4**).

S. N	Types of wastewater	Treatment system	Operating conditions	Pollutants monitored	Removal efficiencies (%)	References
1	Oilfield produced	UASB coupled with immobilized biological aerated filters (I-BAFs)	HRT 12 h (Min)	COD NH$_4^+$-N SS	74 94 98	[39]
		UASB-two stage BAF	Temperature: 26–33°C	COD NH$_4^+$-N Oil PAHs	90.2 90.8 86.5 89.4	[3]
		hydrolysis, MBBR, O$_3$ and biological active carbon reactor		COD Oil Ammonia	95.8 98.9 94.4	[40]
2	Petroleum refinery	MBBR with anaerobic-aerobic (A/O)	HRT: 72 h HRT: 36 h	COD	<60 mg/L (effluent)	[41]
		UASB-aerobic packed bed biofilm reactor (PBBR)	0.5 kg COD/ m^3.d Temperature: 35 ± 1 °C	COD PAHs	81.1 100	[42]
3	Oily wastewater	HyVAB reactor containing anaerobic and aerobic in vertical	To 23 kg COD/m^3.d	COD	86	[9]
		Bioaugmentation anoxic-oxic (A/O)	HRT 17.5 h	COD NH$_4^+$-N	91 89	[5]

Table 4. Overview of integrated treatment process of petrochemical wastewater.

For petrochemical refinery treatment, direct discharge of treatment effluents after combining anaerobic and aerobic MBBR system is possible. The PAH removal reached even 100% by combining the UASB and packed bed biofilm reactor (PBBL) at 0.5 kg COD/m^3·d (**Table 4**).

The pilot study of hybrid vertical flow anaerobic biofilm (HyVAB) treating oily wastewater had substantially high organic loading rate over 23 kg COD/m^3·d. The COD removal efficiency was consistently good over 86% [9]. A case study based on this HyVAB concept is followed in the next section with detailed performance data presentations and discussions.

5. Petrochemical wastewater treatment case study

Petrochemical wastewater of different sources, such as from manufacturing industries, auto repair shops, and washing water of oil tanks, is collected and delivered to a full-scale aerobic treatment plant at Bamble, Norway, for resource recovery and biological stabilization. The collected wastes are stored in storage tanks before being distilled to extract oil residuals. The wastewater after oil extraction still contains high COD and is therefore further treated by biological processes. The full-scale CFIC plant was designed and delivered by Biowater Technology AS and has been running continuously for 3 years. A pilot study of the integrated system HyVAB was also carried out on site of the full-scale plant running with the same feed water and the results showed good performance and can be referred to [9]. In this chapter, the full-scale CFIC operation data and a continuous study of HyVAB applying pure oxygen as aeration media are presented.

5.1. Full-scale CFIC treating petrochemical wastewater

The full-scale plant applies continuous flow intermittent cleaning biofilm (CFIC) technology. The CFIC technology is an advanced biofilm system based on MBBR concept. It is compact and is operated with alternating a normal and a washing mode while continuously feeding the reactor. CFIC contains highly packed biofilm carriers (over 90% filling ratio) to a degree that oxygen is utilized efficiently by enhancing gas transfer and limiting carriers' movement in the reactor. The biofilm grows in condition of sufficient oxygen, organic substrates, and nutrients. Excess aerobic sludge grown on the carriers' surface is washed off during the intermittent washing that helps maintain a thin and effective biofilm.

5.1.1. System layout

The full-scale plant layout is shown in **Figure 1**. Distilled wastewater is pumped to a conditioning chamber where nutrients are dosed and pH is corrected. Effluent from CFIC goes through chemical precipitation and DAF to remove solids before being discharged to the sea. Sludge is temporally stored and dewatered to be tanked away for specific treatment.

The full-scale system is treating wastewater of fluctuating concentrations with COD concentration ranging from 7 to 35 g/L at a designed daily flow rate of 240 m^3/d. The wastewater pH is around 5 and a total dissolved solid content of 4 g/L. BWTS® (Biowater Technology AS) with a surface area of 650 m²/m³ is applied as biofilm carriers in CFIC (**Figure 1**).

5.1.2. Operational results and discussion

Operational data of the full-scale plant in 2017 is summarized here. The COD feed to the reactor and the final effluent after DAF is shown in **Figure 2** together with removal efficiency. It shows that on average over 90% feed COD was removed by the system. At the early days of the year, sludge flocculation process chemical dosing was not well established; the total COD

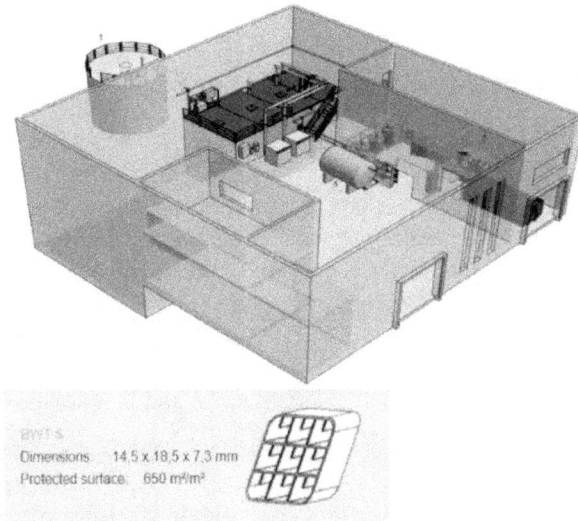

Figure 1. Up, layout of the full-scale CFIC plant with 1. Storage tank; 2. CFIC reactor; 3. DAF; 4. Sludge storage tank; 5. Dewatered sludge tanker. Down, applied BWTS® biofilm carriers.

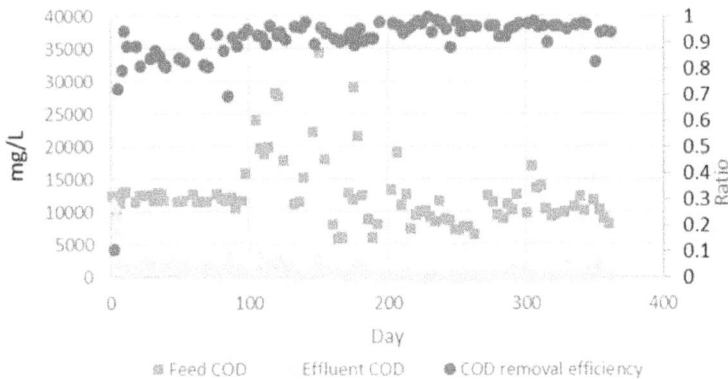

Figure 2. Feed and effluent total COD and COD removal efficiency.

removal was fluctuating around 80–90%. When the system was stabilized even high COD feed from 100 to 200 days did not reduce treatment efficiency. The high removal efficiency indicates that CFIC is a stable and robust system.

The suspended solid content of the final effluent shows that the average value was within 100 mg/L (**Figure 3**). The CFIC system running in normal mode generally worked as a filter bed which retains suspended solid in the reactor. When washing mode starts, raised water level in the reactor coupled with increased aeration induces a well-mixed moving bed biofilm system. The extra biofilm/sludge in carrier voids are washed off due to intensified shear force and are carried out of the system by continuous effluents. The washing washes away on average 30% of the total solids on the biofilm carriers.

5.2. Pilot study of HyVAB treating petrochemical wastewater using pure oxygen

The concept of the HyVAB system is illustrated in **Figure 4**. The system consists of a bottom anaerobic and a top aerobic biofilm stage in a vertical mode. Biogas generated from the anaerobic stage can be collected through the three-stage separator. Due to the close integration of two processes, the dissolved gases (methane, H_2S, etc.) in liquid that are generated in the AD stage will not be released to the atmosphere but captured and oxidized by aerobic organisms, avoiding a commonly observed emission problem in anaerobic treatment plants [43]. Returning of the excess aerobic sludge to the AD stage by gravity where the solids undergo stabilization simplifies the sludge treatment which also contributes to methane production. The detailed longer-term pilot study with reactor layout and performance can be referred to [9], where air was applied as aeration source.

This chapter presents the pilot study of pure oxygen effects on HyVAB performance. Oxygen aerations were known to be less energy intensive, high in efficiency, and give good biofilm development due to its close contact with biofilm layers. Results show that the HyVAB COD removal using air and pure oxygen reached similar ratios on average 94 and 85% for the soluble and total feed COD removal, respectively. Oxygen aeration minimized the flushing

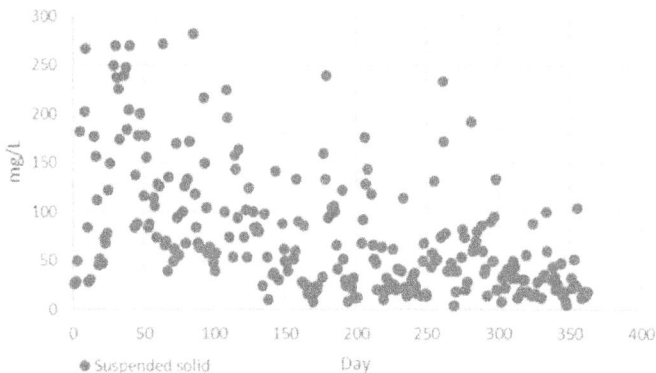

Figure 3. Effluent suspended solid concentration after DAF.

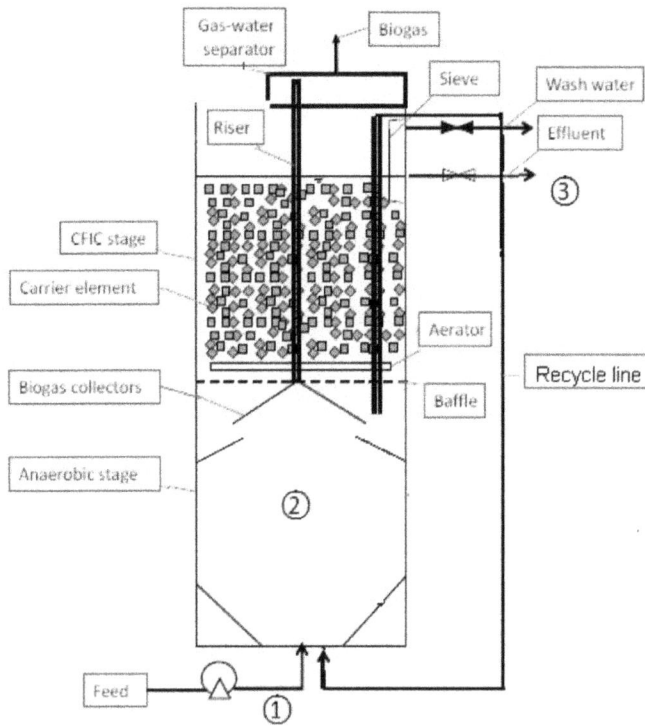

Figure 4. Sketch of the HyVAB (hybrid vertical anaerobic biofilm) bioreactor with the anaerobic stage at the bottom and a CFIC stage at the top. Numbers are sampling points.

effects on biofilm carriers and reduced the effluent suspended solid to 500 mg/L and effluent pH was overall 1.1 less than applying air aeration.

5.2.1. Experiment management

The anaerobic stage was filled with granular sludge, with relatively equal size ($\sim$2 mm) from an industrial wastewater treatment facility. Similar biofilm carriers (**Figure 1**) were used in the aerobic stage. Pure oxygen was applied as aeration oxygen source and air washing was introduced intermittently during the washing mode in the study. The pilot was running continuously for 115 days at 21 $\pm$ 2°C.

5.2.2. Operational results and discussion

With OLR increased gradually to close to 30 kg COD/m³·d at lower HRT of 15 h, the HyVAB system still performed well with over 90% soluble COD removal when the oxygen aeration was introduced after 32 days (**Figure 5**). The air aeration was conducted before 31 days and the results were treated as reference. With oxygen aeration, the anaerobic stage generated high

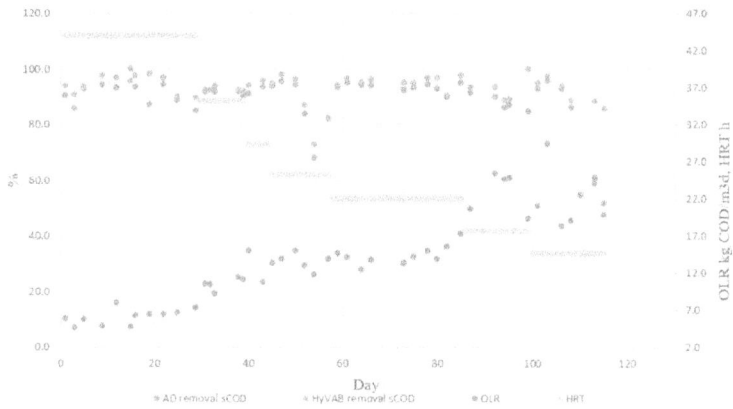

Figure 5. COD removal at different organic loading rate (OLR) and HRT.

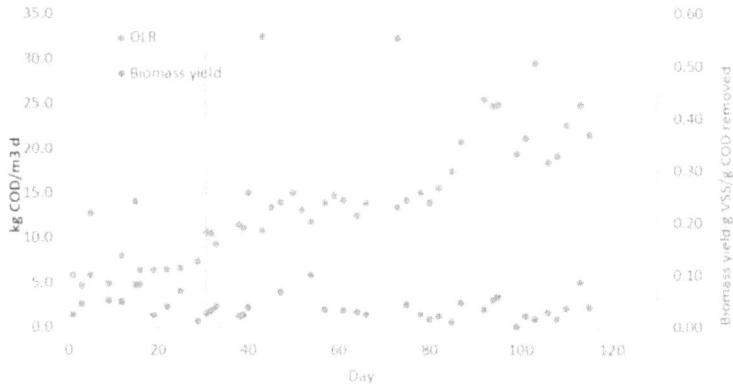

Figure 6. Biomass yield at different OLR and with/without oxygen aeration, vertical line separate air and oxygen aeration.

methane content biogas (82%) and the soluble COD removal efficiency was comparable with air aeration (**Figure 5**).

The sludge yield with oxygen aeration was at 0.04 g VSS/g $COD_{removed}$ and less variations showed comparing to the air aeration stage (**Figure 6**). The reasons can be that the fine bubbles of the aeration from oxygen did not give high shear force on the biofilm to scratch it off. The low mixing effects also retained the solids in the reactor. The low sludge yield at high organic loading rate indicates high efficiency of the HyVAB system in removing petrochemical organic substances. Consistently lower effluents of less than 500 mg/L were observed with oxygen aeration.

Some petrochemical wastewater contains high salinity and nutrients such as ammonia and phosphate, especially after anaerobic treatment. The high content of dissolved solids might

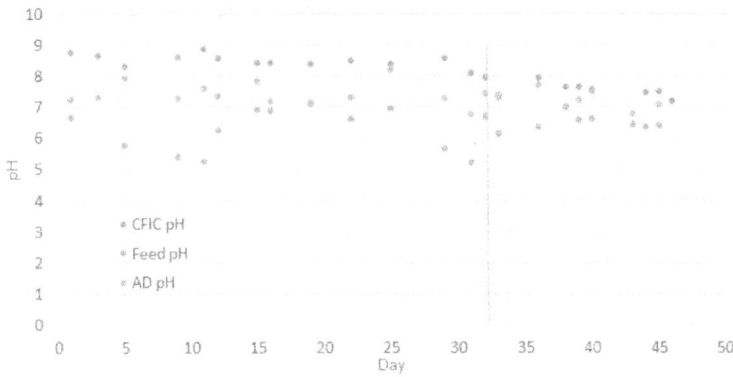

Figure 7. pH variations in different streams, vertical line separate air and oxygen aeration.

precipitate on biofilm carriers when pH is high and temperature is in good range. Oxygen aeration showed good pH control effects compared with air aeration which induced high pH (over 8.5) (**Figure 7**) in the aerobic stage. Good biofilm development was observed in the pilot test with such petrochemical wastewater and the scaling effects on carriers were minimal with oxygen aeration.

6. Conclusions and further study

Biological treatment of petrochemical wastewater is an economic and efficient waste stabilization method. The treatment of wastewater containing organic contaminants of refractory nature can be ineffective in biological treatments [44]. The challenges are as follows: (1) activated sludge method can fail while treating strong petrochemical wastewater with high COD concentration (>10 g/L) and contain some aromatic compounds (phenol and its derivatives, etc.); [45] (2) variations in the strength of the organic load due to various sources of petrochemical refinery can cause shock to the biomass; (3) petrochemical wastewater contains large amounts of volatile organic compounds (VOCs) and can cause odor and air pollutions around the biological treatments, and aerobic treatments like activated sludge should not be considered in this case; (4) oil, fat, and grease can cause floatation of the sludge and this can cause sludge washout ultimately failing the treatment system.

Application of certain organisms for specific wastewater components' treatment after secondary biological treatment can be a topic in the future. Isolation of specific bacteria to treat recalcitrant compounds can lead to effective removal, for example, the bacterium *Pseudomonas putida* to degrade phenolic compounds [7]. Integrated biological system showed general better performance for treating petrochemical wastewater; the synergistic effects of organisms of different originals such as from anaerobic combining aerobic might facilitate the recalcitrant organic removal. Also, reactor modification and microorganisms' isolation to handle complex petroleum wastewater treatment can be of great interest in coming days to reduce extra-treatment costs.

Acknowledgements

The authors would like to thank for the funding provided by Skattefunn No. 265293 and University of South-eastern Norway.

Conflict of interest

There is no conflict of interest.

Notes/Thanks/Other declarations

The authors would like to thank Norsk Spesialolje for supporting this research during pilot study and by providing operational data. Also, our thank goes to University of South-eastern Norway for research cooperation and Praxair for supporting on the pilot study.

Author details

Nirmal Ghimire[1]* and Shuai Wang[2]

*Address all correspondence to: nirmal.ghimire@ku.edu.np

1 University of Southeast Norway, Porsgrunn, Norway

2 Biowater Technology AS, Tønsberg, Norway

References

[1] Gutierrez E, Caldera Y, Fernandez N, Blanco E, Paz N, Marmol Z. Thermophilic anaerobic biodegradability of water from crude oil production in batch reactors. Revista Tecnica de la Facultad de Ingenieria Universidad del Zulia. 2007;**30**:111-117

[2] Llop A, Pocurull E, Borrull F. Evaluation of the removal of pollutants from petrochemical wastewater using a membrane bioreactor treatment plant. Water, Air, and Soil Pollution. 2008;**197**(1–4):349-359

[3] Zou XL. Treatment of heavy oil wastewater by UASB-BAFs using the combination of yeast and bacteria. Environmental Technology. 2015;**36**(18):2381-2389

[4] Tong K, Zhang Y, Liu G, Ye Z, Chu PK. Treatment of heavy oil wastewater by a conventional activated sludge process coupled with an immobilized biological filter. International Biodeterioration & Biodegradation. 2013;**84**:65-71

[5] Zhao X, Wang Y, Ye Z, Borthwick AGL, Ni J. Oil field wastewater treatment in biological aerated filter by immobilized microorganisms. Process Biochemistry. 2006;**41**(7):1475-1483

[6] Fakhru'l-Razi A, Pendashteh A, Abdullah LC, Biak DRA, Madaeni SS, Abidin ZZ. Review of technologies for oil and gas produced water treatment. Journal of Hazardous Materials. 2009;**170**(2–3):530-551

[7] Benyahia F, Abdulkarim M, Embaby A, Rao M. Refinery wastewater treatment: A true technological challenge. In: The Seventh Annual UAE University Research Conference; April 2006. UAE University

[8] Santo CE, Vilar VJP, Bhatnagar A, Kumar E, Botelho CMS, Boaventura RAR. Biological treatment by activated sludge of petroleum refinery wastewaters. Desalination and Water Treatment. 2013;**51**(34–36):6641-6654

[9] Wang S, Ghimire N, Xin G, Janka E, Bakke R. Efficient high strength petrochemical wastewater treatment in a hybrid vertical anaerobic biofilm (HyVAB) reactor: A pilot study. Water Practice Technology. 2017;**12**:501-513

[10] Vincent-Vela MC, Álvarez-Blanco S, Lora-García J, Carbonell-Alcaina C, Sáez Muñoz M. Application of several pretreatment technologies to a wastewater effluent of a petrochemical industry finally treated with reverse osmosis. Desalination and Water Treatment. 2015;**55**(13):3653-3661

[11] Wu C, Zhou Y, Wang P, Guo S. Improving hydrolysis acidification by limited aeration in the pretreatment of petrochemical wastewater. Bioresource Technology. 2015;**194**:256-262

[12] Verma S, Prasad B, Mishra IM. Pretreatment of petrochemical wastewater by coagulation and flocculation and the sludge characteristics. Journal of Hazardous Materials. 2010;**178**(1–3):1055-1064

[13] Lin CK, Tsai TY, Liu JC, Chen MC. Enhanced biodegradation of petrochemical wastewater using ozonation and BAC advanced treatment system. Water Research. 2001;**35**(3):699-704

[14] Li G, Guo S, Li F. Treatment of oilfield produced water by anaerobic process coupled with micro-electrolysis. Journal of Environmental Sciences. 2010;**22**(12):1875-1882

[15] Dai X, Chen C, Yan G, Chen Y, Guo S. A comprehensive evaluation of re-circulated biofilter as a pretreatment process for petroleum refinery wastewater. Journal of Environmental Sciences. 2016;**50**:49-55

[16] Jamaly S, Giwa A, Hasan SW. Recent improvements in oily wastewater treatment: Progress, challenges, and future opportunities. Journal of Environmental Sciences (China). 2015;**37**:15-30

[17] Diaz A, Rincon N, Martin J, Behling E, Chacin E, Debellefontaine H. Degradation of total phenol during biological treatment of oilfields produced water. Ciencia. 2005;**13**:281-291

[18] Gutierrez E, Caldera Y, Perez F, Blanco E, Paz N. Behavior of metals in water from crude oil production during thermophilic anaerobic treatment. Boletin del Centro de Investigaciones Biologicas. 2009;**43**(1):145-160

[19] Rincon N, Cepeda N, Diaz A, Behling E, Marin JC, Bauza R. Behavior of organic fraction in water separated from extracted crude oil with anaerobic digestion. Revista Tecnica de la Facultad de Ingenieria Universidad del Zulia. 2008;**31**:169-176

[20] Rincon N, Chacin E, Marin J, Moscoso J, Fernandez L, Torrijos M, Moletta R, Fernandez N. Optimum time of hydraulic retention for the anaerobic treatment of light oil production wastewater. Revista Tecnica de la Facultad de Ingenieria Universidad del Zulia. 2002;**25**:90-99

[21] Gutierrez E, Caldera Y, Contreras K, Blanco E, Paz N. Anaerobic mesophilic and thermophilic degradation of waters from light crude oil production. Boletin del Centro de Investigaciones Biologicas. 2006;**40**:242-256

[22] Caldera Y, Gutierrez E, Madueno P, Griborio A, Fernandez N. Anaerobic biodegradability of industrial effluents in a UASB reactor. Impacto Cientifico. 2007;**2**:11/23

[23] Gasim HA, Kutty SRM, Isa MH, Alemu LT. Optimization of anaerobic treatment of petroleum refinery wastewater using artificial neural networks. Research Journal of Applied Sciences, Engineering and Technology. 2013;**6**:2077-2082

[24] Ji GD, Sun TH, Ni JR, Tong JJ. Anaerobic baffled reactor (ABR) for treating heavy oil produced water with high concentrations of salt and poor nutrient. Bioresource Technology. 2009;**100**:1108-1114

[25] Mirbagheri SA, Ebrahimi M, Mohammadi M. Optimization method for the treatment of Tehran petroleum refinery wastewater using activated sludge contact stabilization process. Desalination and Water Treatment. 2013;**52**:156-163

[26] Thakur C, Srivastava VC, Mall ID. Aerobic degradation of petroleum refinery wastewater in sequential batch reactor. Journal of Environmental Science and Health. Part A, Toxic/Hazardous Substances & Environmental Engineering. 2014;**49**:1436-1444

[27] Pajoumshariati S, Zare N, Bonakdarpour B. Considering membrane sequencing batch reactors for the biological treatment of petroleum refinery wastewaters. Journal of Membrane Science. 2017;**523**:542-550

[28] Razavi SMR, Miri T. A real petroleum refinery wastewater treatment using hollow fiber membrane bioreactor (HF-MBR). Journal of Water Process Engineering. 2015;**8**:136-141

[29] Rahman MM, Al-Malack MH. Performance of a crossflow membrane bioreactor (CF–MBR) when treating refinery wastewater. Desalination. 2006;**191**(1–3):16-26

[30] Xie W, Zhong L, Chen J. Treatment of slightly polluted wastewater in an oil refinery using a biological aerated filter process. Wuhan University Journal of Natural Sciences. 2007;**12**:1094-1098

[31] Vendramel S, Bassin JP, Dezotti M, Sant'Anna GL Jr. Treatment of petroleum refinery wastewater containing heavily polluting substances in an aerobic submerged fixed-bed reactor. Environmental Technology. 2015;**36**:2052-2059

[32] Dong Z, Lu M, Huang W, Xu X. Treatment of oilfield wastewater in moving bed biofilm reactors using a novel suspended ceramic biocarrier. Journal of Hazardous Materials. 2011;**196**:123-130

[33] Tellez T, Gilbert NK, Gardea-Torresdey J. Performance evaluation of an activated sludge system for removing petroleum hydrocarbons from oilfield produced water. Advances in Environmental Research. 2002;**6**:455-470

[34] Campos JC, Borges RMH, Oliveira Filho AM, Nobrega R, Sant'Anna GL Jr. Oilfield wastewater treatment by combined microfiltration and biological processes. Water Research. 2002;**36**(1):95-104

[35] Shokrollahzadeh S, Azizmohseni F, Golmohammad F, Shokouhi H, Khademhaghighat F. Biodegradation potential and bacterial diversity of a petrochemical wastewater treatment plant in Iran. Bioresource Technology. 2008;**99**:6127-6133

[36] Ma F, Guo JB, Zhao LJ, Chang CC, Cui D. Application of bioaugmentation to improve the activated sludge system into the contact oxidation system treating petrochemical waste-water. Bioresource Technology. 2009;**100**(2):597-602

[37] Scholz W, Fuchs W. Treatment of oil contamintaed wastewater in a membrane bioreactor. Water Research. 2000;**34**:3621-3629

[38] Chavan A, Mukherji S. Treatment of hydrocarbon-rich wastewater using oil degrading bacteria and phototrophic microorganisms in rotating biological contactor: Effect of N:P ratio. Journal of Hazardous Materials. 2008;**154**:63-72

[39] Liu G-h, Ye Z, Tong K, Zhang Y-h. Biotreatment of heavy oil wastewater by combined upflow anaerobic sludge blanket and immobilized biological aerated filter in a pilot-scale test. Biochemical Engineering Journal. 2013;**72**:48-53

[40] Zheng T. A compact process for treating oilfield wastewater by combining hydrolysis acidification, moving bed biofilm, ozonation and biologically activated carbon techniques. Environmental Technology. 2016;**37**:1171-1178

[41] Lu M, Gu LP, Xu WH. Treatment of petroleum refinery wastewater using a sequential anaerobic-aerobic moving-bed biofilm reactor based on suspended ceramsite. Water Science and Technology. 2013;**67**:1976-1983

[42] Nasirpour N, Mohammad Mousavi S, Shojaosadati SA. Biodegradation potential of hydrocarbons in petroleum refinery effluents using a continuous anaerobic-aerobic hybrid system. Korean Journal of Chemical Engineering. 2015;**32**:874-881

[43] Daelman MR, van Voorthuizen EM, van Dongen UG, Volcke EI, van Loosdrecht MC. Methane emission during municipal wastewater treatment. Water Research. 2012;**46**(11): 3657-3670

[44] Bahri M, Mahdavi A, Mirzaei A, Mansouri A, Haghighat F. Integrated oxidation process and biological treatment for highly concentrated petrochemical effluents: A review. Chemical Engineering and Processing Process Intensification. 2018;**125**:183-196. DOI: 10.1016/j.cep.2018.02.002

[45] Debellefontaine H, Chakchouk M, Foussard J, Tissot D, Striolo P. Treatment of organic aqueous wastes: Wet air oxidation and wet peroxide oxidation. Environmental Pollution. 1996;**92**(2):155-164

www.ingramcontent.com/pod-product-compliance
Lightning Source LLC
Chambersburg PA
CBHW081241190326

41458CB00016B/5874